Lecture Notes in Mathematics

Edited by A. Dold and B. Eckmann

483

Robert D. M. Accola

Riemann Surfaces, Theta Functions, and Abelian Automorphisms Groups

Springer-Verlag
Berlin · Heidelberg · New York 1975

Author
Prof. Robert D. M. Accola
Department of Mathematics
Brown University
Providence, R.I. 02912
USA

Library of Congress Cataloging in Publication Data

Accola, Robert D M 1929-
Riemann sufaces, theta functions, and abelian automorphism groups.

(Lecture notes in mathematics ; 483)
Bibliography: p.
Includes index.
1. Riemann surfaces. 2. Functions, Theta. 3. Automorphisms. 4. Abelian groups. I. Title. II. Series: Lecture notes in mathematics (Berlin) ; 483.
QA3.L28 no. 483 [QA333] 510'.8 [515'.223] 75-25928

AMS Subject Classifications (1970): 14H40, 30A46

ISBN 3-540-07398-1 Springer-Verlag Berlin · Heidelberg · New York
ISBN 0-387-07398-1 Springer-Verlag New York · Heidelberg · Berlin

Printed in Germany
Offsetdruck: Julius Beltz, Hemsbach/Bergstr.

Contents

Part I

1 Introduction 1

2 Remarks on general coverings 4

3 Resumé of the Riemann vanishing theorem 7

4 Ramified normal coverings 8

5 Abelian covers 12

6 Main results 19

Part II

1 Introduction 32

2 Completely ramified abelian covers 40

3 Two-sheeted covers 50

4 Other applications 56

5 Closing remarks 63

Part III

1 Introduction 66

2 Castelnuovo's method and p_0-hyperellipticity 70

3 Extensions 74

4 The p - 2 conjecture for p = 5 79

5 Elliptic-hyperelliptic surfaces of genus five 81

6 Elliptic-hyperelliptic surfaces of genus three 88

7 Cyclic groups of order three for genus two 94

8 Some local characterizations 95

9 Closing remarks 98

References 100

Index 102

PART I[1)]

1. <u>Introduction</u>. Torelli's theorem states that the conformal type of a Riemann surface is determined by (the equivalence class of) one of its period matrices. If a Riemann surface has some property not shared by all Riemann surfaces then this fact should be characterized by some property of the period matrix, hopefully a property which is independent of the particular period matrix at hand. The main tool for effecting such characterizations is Riemann's solution to the Jacobi inversion problem, often called Riemann's vanishing theorem. Riemann's theorem relates vanishing properties of the theta function for the Jacobian of a surface to certain linear series on the surface. Since special properties on a Riemann surface often imply the existence of special linear series, these special properties will be reflected, via Riemann's theorem, in special vanishing properties of the theta function.

---

1)The research for this paper has been carried on during the last several years during which the author received support from several sources. i) Research partially sponsored by the Air Force Office of Scientific Research, Office of Aerospace Research, United States Air Force, under AFOSR Grant No. AF-AFOSR-1199-67. ii) National Science Foundation Grant GP-7651. iii) Institute for Advanced Study Grant-In-Aid.

The special property that a surface might possess considered in this paper is the existence of an abelian group of automorphisms. This subject has a long history. The vanishing properties of hyperelliptic theta functions have been known since the last century [15] Recently Farkas [9] discovered special vanishing properties for theta functions associated with Riemann surfaces which admit fixed point free automorphisms of period two. The author has discovered other vanishing properties for some surfaces admitting abelian automorphism groups of low order. The purpose of this paper is to present a general theory which will include most of the known results.

Part I of this paper will concern the general theorem on vanishing properties of theta functions for surfaces admitting an arbitrary abelian group of automorphisms. Part II will be concerned with applications to particular situations where the order of the group is small. The case where the order of the group is two will be considered in some detail.[2)]

Part III will deal with the problem of the extent to which special vanishing properties characterize surfaces admitting

---

2) The problem of vanishing properties of theta functions for surfaces admitting automorphisms of period two dates from the nineteenth century. The unramified case was treated by Riemann [21, Nachtrage p. 108] Schottky-Jung [24] and more recently by Farkas [9],[10] and Farkas-Rauch [12]. For the hyperelliptic case Krazer [15] has a complete treatment. The elliptic-hyperelliptic case was treated by Roth [22]. Recently the general ramified case been treated extensively by Fay [13]. The above remarks are by no means a complete bibliography. For further references to the work of the nineteenth century the reader is referred to the article by Krazer-Wirtinger [16] and Krazer [15]. The papers of Farkas and Farkas-Rauch contain further references to more modern work.

abelian groups of automorphisms. This seems to be a more difficult problem for which no general theory presently exists.

The setting for these results is a closed Riemann surface, W_1, of genus p_1, $p_1 \geq 2$, admitting a finite abelian group of automorphisms, G. The space of orbits of G, W/G $(=W_0)$, is naturally a Riemann surface so that the quotient map, $\underline{b}$, is analytic. We will, however, consider some results in a more arbitrary setting where the cover $\underline{b} : W_1 \to W_0$ need not be normal.

2. Remarks on General Coverings. Let $\underline{b} : W_1 \rightarrow W_0$ be an arbitrary n-sheeted ramified covering of closed Riemann surfaces of genera p_1 and p_0 respectively. Let M_1 be the field of meromorphic functions on W_1 and let M_0 be the lifts, via $\underline{b}$, of the field of meromorphic functions on W_0. Then M_0 is a subfield of M_1 of index n. We now define an important abelian group, A, as follows:

Definition: $A = \{f \in M_1^* | f^n \in M_0^*\}/M_0^*$. [3)]

Now let M_A be the maximal abelian extension of M_0 in M_1.

Lemma 1: A is isomorphic to the (dual of the) Galois group of M_A over M_0.

Proof: (omitted). A proof in the case where $M_1 = M_A$ will follow in Section 5.

Now, fix a point W_1, z_1, and let $z_0 = \underline{b}(z_1)$. Fix canonical homology bases in W_1 and W_0 and choose bases for the analytic differentials dual to these homology bases. Thus maps u_1 and u_0 from W_1 and W_0 into their Jacobians, $J(W_1)$ and $J(W_0)$, are defined: [4)]

---

3) If M is a field, M^* will stand for the multiplicative group of non-zero elements of M.

4) If $p_0 = 0$ $J(W_0)$ will be taken to be the one element abelian group and the theta function will be the function which takes the value one on $J(W_0)$.

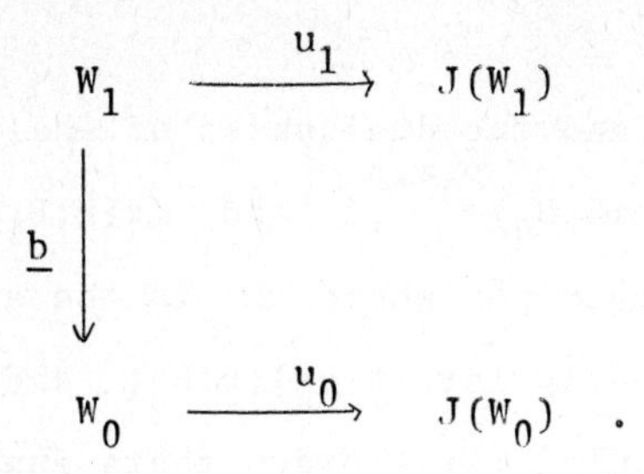

The maps u_1 and u_0 are extended to divisors in the usual way. A map $\underline{a}$ [5] is now defined from divisors on W_0 to those of W_1 as follows: for $x_0 \in W_0$, $\underline{a}x_0$ is the inverse image of x_0 under $\underline{b}$ with branch points counted according to multiplicity. Thus $\underline{a}x_0$ always has degree n. $\underline{a}$ is extended by linearity to arbitrary divisors on W_0. Now we define a map from $J(W_0) \to J(W_1)$, again denoted by $\underline{a}$ as follows: if D_0 is a divisor on W_0 of degree zero, then $\underline{a}u_0(D_0) = u_1(\underline{a}D_0)$. $\underline{a}$ is easily seen to be a homomorphism. Let M_{UA} be the maximal unramified abelian extension of M_0 in M_1; thus $M_0 \subset M_{UA} \subset M_A \subset M_1$.

<u>Lemma 2</u>: The kernel of $\underline{a} : J(W_0) \to J(W_1)$ is isomorphic to the Galois group of M_{UA} over M_0.

<u>Proof</u>: (omitted). A proof in the case when $M_{UA} = M_1$ will follow in Section 5 of Part I. The case $M_A = M_1$ will be

---

5) The symbol $\underline{a}$ will be used consistently to denote homomorphism from abelian groups associated with W_0 into the corresponding abelian groups associated with W_1. The particular group will be clear from the context.

covered in Part II.

With the homology bases and the dual bases of analytic differentials chosen, let $(\pi iE;B_0)^{p_0\times 2p_0}$ and $(\pi iE;B_1)^{p_1\times 2p_1}$ be the corresponding period matrices where E is the appropriate identity matrix. Finally let $\theta[\chi_0](u;B_0)$ and $\theta[\chi_1](u;B_1)$ be the corresponding first order theta functions with arbitrary characteristics.

Lemma 3: For any characteristic χ_1, there is an exponential function $E(u)$ so that $E(u)\theta[\chi_1](\underline{a}u;B_1)$, as a multiplicative function on $J(W_0)$, is an nth order theta function.

Proof: The proof is an immediate adaptation of the simplest parts of transformation theory. Since the technics used in the proof of this lemma are quite different from those used elsewhere in Part I, the proof is deferred until Part II.

3. Resumé of the Riemann Vanishing Theorem. The proofs of the results in this paper depend on Riemann's solution to the Jacobi inversion problem. We summarize here those portions of the theory that will be needed later. [6)]

Let W be a closed Riemann surface of genus p, $p \geq 1$, let a canonical homology basis be chosen, let a dual basis of analytic differenitals be chosen, let a base point be chosen, and let u be the map of W into $J(W)$. Riemann's theorem asserts the existence of a point K in $J(W)$ so that if we choose any $e \in J(W)$, then there is a integral divisor D on W of degree p so that

$$u(D) + K \equiv e \pmod{J(W)}.$$

If $\theta(e) \neq 0$, then D is unique. If $\theta(e) = 0$, then the above equation can be solved with an integral divisor of degree $p - 1$. Moreover, in this latter case, the order of vanishing of $\theta(u)$ at e equals $i(D)$, the index of speciality of D. (By the Riemann-Roch theorem, $i(D)$ equals the number of linearly independent meromorphic functions which are multiples of $-D$ since the degree of D is $p - 1$.) Moreover, $\theta(u(D)) + K) = 0$ whenever D is an integral divisor of degree at most $p - 1$. Finally, if D is a canonical divisor, then

$$u(D) \equiv -2K \pmod{J(W)}.$$

---

6) The material in this section is covered in Krazer [15]. For a complete and more modern treatment see Lewittes [18].

4. Ramified Normal Coverings. Let

$$\underline{b} : W_1 \to W_0$$

be an n-sheeted (possibly) ramified normal covering where the group of automorphisms, G, of W_1 need not be abelian. Let $x_{01}, x_{02}, \dots, x_{0s}$, $s \geq 0$, be the points of W_0 over which the ramification occurs. Let X_0 be the divisor $x_{01} + x_{02} + \cdots + x_{0s}$. For each x_{0j} there is an integer, ν_j, so that above x_{0j} there are n/ν_j branch points each of multiplicity ν_j, $j = 1,2, \dots, s$. The Riemann-Hurwitz formula for this cover is

$$(1) \quad 2p_1 - 2 = n(2p_0 - 2) + r$$

where the total ramification, r, is given by

$$(2) \quad r = n \sum_{j=1}^{s} (1 - \nu_j^{-1})$$

If $\underline{b}$ is unramified $(r = 0)$ then most of the following discussion is unnecessary.

Let x_{1j} be the inverse image of x_{0j} under $\underline{b}$ considered as a divisor of degree n/ν_j. Let X_1 be the divisor on W_1 $x_{11} + x_{12} + \cdots + x_{1s}$. Then

$$\underline{a}x_{0j} = \nu_j x_{1j}$$

and $$\underline{a}u_0(x_{0j} - z_0) \equiv u_1(\underline{a}x_{0j} - \underline{a}z_0).$$

So (3) $\underline{a}u_0(x_{0j}) \equiv u_1(\underline{a}x_{oj}) - u_1(\underline{a}z_0)$

There are $\nu_j^{2p_0}$ points on $J(W_0)$ which when multiplied by ν_j give $u_0(x_{0j})$ for $j = 1,2, \ldots ,s$. Fix one of these and denote it by $\nu_j^{-1}u_0(x_{0j})$. Similarly, let $n^{-1}u_1(\underline{a}z_0)$ denote a certain point of $J(W_1)$ which when multiplied by n gives $u_1(\underline{a}z_0)$. Let $\nu_j^{-1}u_1(\underline{a}z_0) = (n/\nu_j)n^{-1}u_1(\underline{a}z_0)$. Then dividing formula (3) by ν_j and rearranging terms yields

(4) $u_1(x_{1j}) \equiv \underline{a}(\nu_j^{-1}u_0(x_{0j})) + \nu_j^{-1}u_1(\underline{a}z_0) + c_j$

where $\nu_j c_j \equiv 0$. c_j depends on the choice of $n^{-1}u_1(\underline{a}z_0)$ and $\nu_j^{-1}u_0(x_{0j})$.

Now we determine K_1, the vector of Riemann constants on W_1, in terms of $\underline{a}K_0$ and other quantities already defined. Let D_0 be an integral canonical divisor on W_0. Then

$$\underline{a}D_0 + \sum_{j=1}^{s}(\nu_j - 1)x_{1j}$$

a divisor of degree $n(2p_0 - 2) + r(= 2p_1 - 2)$, is canonical on W_1. Now

$$-2K_0 \equiv u_0(D_0 - (2p_0 - 2)z_0)$$

So $$-2\underline{a}K_0 \equiv u_1(\underline{a}D_0) - (2p_0 - 2)u_1(\underline{a}z_0).$$

Since $$-2K_1 \equiv u_1(\underline{a}D_0) + \sum_{j=1}^{s}(\nu_j - 1)u_1(x_{1j})$$

we have

$$-2K_1 \equiv -2\underline{a}K_0 + (2p_0 - 2)u_1(\underline{a}z_0) + \sum_{j=1}^{s}(\nu_j - 1)u_1(x_{ij}).$$

Substitute formula (4) into this last equation.

$$-2K_1 \equiv -2\underline{a}K_0 + (2p_0 - 2)u_1(\underline{a}z_0) +$$

$$\sum_{j=1}^{s}(\nu_j - 1)\left[\underline{a}(\nu_j^{-1}u_0(x_{0j})) + \nu_j^{-1}u_1(\underline{a}z_0) + c_j\right]$$

Noting that $\nu_j c_j \equiv 0$ and $2p_0 - 2 + \sum_{j=1}^{s}(\nu_j - 1)\nu_j^{-1} = n^{-1}(2p_1 - 2)$, and dividing this last equation by two gives

(5) $\quad K_1 \equiv \underline{a}K_0 - (p_1 - 1)n^{-1}u_1(\underline{a}z_0) - \underline{a}e_0 - e_1$

where e_0 and e_1 are points of $J(W_0)$ and $J(W_1)$ respectively so that

(6) $\quad 2e_0 \equiv \sum_{j=1}^{s}(\nu_j - 1)\nu_j^{-1}u_0(x_{0j})$ [7]

(7) $\quad 2e_1 \equiv -\sum_{j=1}^{s}c_j.$

Thus $2ne_1 \equiv 0$.

If the cover $\underline{b} : W_1 \to W_0$ is unramified, we have simply the equation

---

7) For future references we define e_0 so that $ne_0 = \sum(n(\nu_j - 1)/2)\nu_j^{-1}u_0(x_{oj})$. If n is even this will always be true. If n is odd so are all ν_j so that if n is odd we define $e_0 = \sum_{j=1}^{s}((\nu_j - 1)/2)\nu_j^{-1}u_0(x_{0j})$.

(8) $K_1 \equiv \underline{a}K_0 - (p_0-1)u_1(\underline{a}z_0) - e_1$

where $2e_1 \equiv 0$, since $p_0 - 1 = (p_1-1)n^{-1}$ in this case. e_0 is taken to be zero in this case.

5. Abelian Covers.[8] In this section assume $\underline{b} : W_1 \to W_0$ is a (possibly) ramified abelian cover; i.e., $W_0 = W_1/G$ where G is an abelian group of automorphisms of W_1 whose elements may have fixed points. Let R be the set of characters of G; i.e., the set of homomorphisms of G into the multiplicative group of complex numbers of modulus one. Since G is a finite group, R is isomorphic to G although there is no canonical isomorphism. The group A is, however, canonically isomorphic to R. For suppose f is meromorphic on W_1 and f^n is a function lifted from W_0. Then the divisor of f is invariant under G. Thus, if $T \in G$, then $f \circ T = \chi_f(T) f$ where χ_f is easily seen to be a character. If f and g yield the same character, then f/g gives the identity character and so lies in M_0. Thus the map $f \to \chi_f$ is an isomorphism of A into R. That this map is onto is seen by examining, for each $\chi \in R$, the cyclic extension of M_0 given by the fixed field for kernel χ. This completes a proof of Lemma 1 when $M_1 = M_A$. In the above situation where we have a function f whose divisor is invariant under G, we shall say that f <u>corresponds</u> to the character χ_f.

Thus the field extension M_1 over M_0 can have as a vector space basis functions corresponding to each character of R. If f if an arbitrary function, then let $f_\chi = n^{-1} \sum_{T \in G} \chi(T^{-1}) f \circ T$.

---

8) The author wishes to express his thanks to Professor M. S. Narasimhan for many valuable discussions concerning the material of this paper, especially this section.

Then $f = \sum_{\chi \in R} f_\chi$ and the f_χ's which are not zero are linearly independent since they correspond to different characters.

The next part of this section generalizes to the context of this paper the discussion of half-periods that procedes the classical statements of the vanishing properties of the hyperelliptic theta functions. If the cover is unramified then this discussion is unnecessary.

Lemma 4: Let α be a function in M_1 so that $\alpha^n \in M_0$. Then there is a divisor $D_{0\alpha}$ on W_0 not containing any point of X_0 and an s-tuple of integers $(\alpha_1, \alpha_2, \ldots, \alpha_s)$ so that

$$(9) \qquad (\alpha) = \underline{a} D_{0\alpha} + \sum_{j=1}^{s} \alpha_j x_{1j}$$

Moreover

$$(10) \qquad nD_{0\alpha} + \sum_{j=1}^{s} (n/\nu_j)\alpha_j x_{0j} \equiv 0 \qquad \text{on } W_0.$$

If α' has the same properties as α, α' corresponds to the same character as α, and $(\alpha_1', \alpha_2', \ldots, \alpha_s')$ is the s-tuple for α' then for $j = 1, 2, \ldots, s$

$$\alpha_j \equiv \alpha_j' \pmod{\nu_j}$$

Proof: Since (α) is invariant under G, it must be of the given form. α^n is a function lifted via $\underline{b}$ from W_0 (call this function on W_0, α_0) and the divisor of α_0 must be the left hand side of formula (10). For α' with the given properties it follows that α/α' corresponds to the identity character; that is, α/α' itself is in M_0 Thus $\alpha_j - \alpha_j' \equiv 0 \pmod{\nu_j}$ for all j. q.e.d.

Lemma 4 shows that the s-tuple $(\alpha_1, \alpha_2, \dots, \alpha_s)$ viewed as a point in $Z_{\nu_1} \times Z_{\nu_2} \times \cdots \times Z_{\nu_s}$ (call this abelian group V) depends only on the character, χ, to which α corresponds. Let χ_j be the smallest non-negative residue of $\alpha_j \pmod{\nu_j}$ and consider the map $R \to V$ given by

$$\chi \to (\chi_1, \chi_2, \dots, \chi_s).$$

This map is a homomorphism. The hyperelliptic case shows that it need not be onto. Since an α which generates an unramified cyclic extension gives the zero s-tuple, the homomorphism need not be one to one. There are, however, properties of this homomorphism which we will need later.

Lemma 5: For a given j, there is a $\chi \in R$ so that $\chi_j = 1$.

Proof. Let $\langle T\rangle$ be the stabilizer of one (and therefore all) of the points of x_{1j}. Call the point q. Then $T^{\nu_j} = \mathrm{id}$. There exists a character χ which is faithful on the cyclic group $\langle T\rangle$. Thus there is a function, α, corresponding to χ so that

$$\alpha \circ T = \omega\alpha$$

where $\omega = \mathrm{Exp}\{2\pi i/\nu_j\}$. We can find a local parameter, z, at q so that $z(q) = 0$ and T has the representation

$Tz = \omega z$. Thus in this parameter

(11) $\alpha(\omega z) = \omega\alpha(z)$

Now $\alpha(z) = \sum_{k=t}^{\infty} \lambda_k z^k$ near q for some coefficients λ_k. Formula (11) implies that $\lambda_j = 0$ unless $j \equiv 1 \pmod{\nu_j}$. Consequently $\alpha_j \equiv 1 \pmod{\nu_j}$ and so $\chi_j = 1$. q.e.d.

Lemma 6: For any $\chi \in R$ we have

(12) $\sum_{j=1}^{s} (n/\nu_j)\chi_j \equiv 0 \pmod{n}$

Proof: By Lemma 4, formula (10)

$$n \deg D_{0\alpha} + \sum_{j=1}^{s} (n/\nu_j)\alpha_j = 0$$

since $\deg x_{0j} = 1$. Since $\chi_j \equiv \alpha_j \pmod{\nu_j}$ the result follows.

q.e.d.

Now consider $\beta = (\beta_1, \beta_2, \ldots, \beta_s) \in V$ where $0 \le \beta_j < \nu_j$ for each j. Assume the β_j's are chosen so that

(13) $\sum_{j=1}^{s} (n/\nu_j)\beta_j \equiv r/2 \pmod{n}$.

where r is the ramification. For $\chi \in R$ define $\beta_{\chi j}$ by

(14) $\beta_{\chi j} \equiv \beta_j + \chi_j \pmod{\nu_j}$

where $0 \le \beta_{\chi j} < \nu_j$. For $\chi \in R$ define t_χ by the following formula

$$(15) \quad \sum_{j=1}^{s} (n/\nu_j)\beta_{\chi j} = (r/2) - t_\chi n.$$

By Lemma 6 this is possible.

<u>Lemma 7</u>. Given $\beta \in V$ satisfying formula (14) we have

$$(16) \quad \sum_{\chi \in R} t_\chi = 0$$

<u>Proof</u>. $\sum_R \sum_j (n/\nu_j)\beta_{\chi j} = \sum_j (n/\nu_j) \sum_R \beta_{\chi j}$

Fix j. Since $\chi_j = 1$ for some χ, χ_j runs through the numbers $0,1,2,\ldots,\nu_j - 1$ (n/ν_j) times as χ runs through R. The same is true for $\beta_{\chi j}$. Consequently

$$\sum_R \beta_{\chi j} = (n/\nu_j)((\nu_j^2 - \nu_j)/2) = n(\nu_j - 1)/2$$

Thus $\sum_R \sum_j (n/\nu_j)\beta_{\chi j} = \sum_j (n/\nu_j)(n(\nu_j - 1)/2) = (nr)/2$

By formula (15) $\sum_R ((r/2) - t_\chi n) = (nr)/2$.
The result follows. q.e.d.

The next part of this section examines more closely the divisor in formula (10).

By Lemma 4, formula (10)

$$u_0(nD_0 + \sum_{j=1}^{s} (n/\nu_j)\alpha_j x_{0j}) \equiv 0$$

Thus there is a (1/n)-period in $J(W_0)$, provisionally called ε_α, (since it seems to depend on α) so that

$$(17) \quad u_0(D_{0\alpha}) + \sum_{j=1}^{s} \alpha_j \nu_j^{-1} u_0(x_{0j}) \equiv \varepsilon_\alpha$$

If α and α' correspond to the same character then

$$D_{0\alpha} - D_{0\alpha}' + \sum((\alpha_j - \alpha_j')/\nu_j)x_{0j} \equiv 0$$

Thus $\varepsilon_\alpha \equiv \varepsilon_{\alpha'}$. Consequently, ε_α depends only on the character to which α corresponds, so we denote it ε_χ. The map $\chi \to \varepsilon_\chi$ from $R \to J(W_0)$ is clearly a homomorphism. Applying $\underline{a}$ to formula (17) and letting $d = \deg D_{0\alpha}$ gives

$$\underline{a}u_0(D_{0\alpha} - dz_0) + \sum \alpha_j \underline{a}(\nu_j^{-1} u_0(x_{0j})) \equiv \underline{a}\varepsilon_\alpha .$$

Applying formula (4) gives

$$u_1(\underline{a}D_{0\alpha}) - du_1(\underline{a}z_0) +$$

$$\sum \alpha_j [u_1(x_{1j}) - \nu_j^{-1} u_1(\underline{a}z_0) - c_j] \equiv \underline{a}\varepsilon_\alpha$$

But $\underline{a}D_{0\alpha} + \sum \alpha_j x_{1j} \equiv 0$. Consequently $nd + \sum \alpha_j (n/\nu_j) = 0$, and so $d + \sum(\alpha_j/\nu_j) = 0$. Since $\nu_j c_j \equiv 0$ and $\chi_j \equiv \alpha_j \pmod{\nu_j}$ we obtain the following lemma.

<u>Lemma 8</u>. (18) $\underline{a}\varepsilon_\chi \equiv -\sum_{j=1}^{s} \chi_j c_j$

We conclude this section by giving a proof of Lemma 2 in the case where $\underline{b} : W_1 \to W_0$ is an unramified abelian cover. In this case Lemma 8 shows that the map $\chi \to \varepsilon_\chi$ is a homomorphism from R into $\ker \underline{a}$. If $\varepsilon_\chi = \varepsilon_{\chi'}$, then

$u_0(D_{0\alpha}) \equiv u_0(D_{0\alpha'})$ and so α/α' corresponds to the identity character. Thus $\chi = \chi'$ and the map is one to one. To show that the map is onto, suppose $\underline{a}\varepsilon \equiv 0$. If $u_0(D_0) \equiv \varepsilon$ where $\deg D_0 = 0$, then $\underline{a}u_0(D_0) \equiv u_1(\underline{a}D_0) \equiv 0$. Thus $\underline{a}D_0$ is principal and invariant under G. If $\underline{a}D_0 = (\alpha)$ then α corresponds to some character χ, and $\varepsilon = \varepsilon_\chi$.

6. Main Results. The statement and proof of the main result of this paper are technically complicated. However, the idea of the proof is exactly the same as in the unramified case.[9] In the statement of the proof some of the previous notation is summarized, but not all.

At this point let us introduce a convenient abuse of notation. The reader is reminded that throughout this discussion canonical homology bases on W_0 and W_1 are fixed and all results are related to these bases. If $\theta(u;B)$ is a theta function and $\tau = \pi i\tilde{h} + B\tilde{g}$ where g and h are real row vectors, then there is an exponential function $E(u)$ so that

$$\theta(u + \tau;B) = E(u)\theta\begin{bmatrix} g \\ h \end{bmatrix}(u;B).$$

In this context write $\theta[\tau](u)$ for the usual $\theta\begin{bmatrix} g \\ h \end{bmatrix}(u;B)$. This notation will be extremely convenient and will lead to no confusion provided the canonical homology bases remain fixed.

Theorem 1: Fix s integers $\beta_1, \beta_2, \ldots, \beta_s$ so that

$$0 \le \beta_j < \nu_j \quad \text{for } j = 1,2, \ldots, s$$

and so that $\sum_{j=1}^{s} (n/\nu_j)\beta_j - r/2$ is an integer which is a multiple of n. For each χ in R define $\beta_{\chi j}$ by

---

9) See Accola [4].

(19) $\beta_{\chi j} \equiv \beta_j + \chi_j \pmod{\nu_j}$

and $0 \leq \beta_{\chi j} < \nu_j$. Then define t_χ by the equation

(20) $\sum_{j=1}^{s} (n/\nu_j)\beta_{\chi j} = r/2 - t_\chi n.$

Fix $g_0 \in J(W_0)$. For each $\chi \in R$ consider the equation

(21) $g_0 + e_0 - \sum_{j=1}^{s} \beta_{\chi j}\nu_j^{-1}u_0(x_{0j}) + \varepsilon_\chi \equiv u_0(\sigma_\chi) + K_0.$

Define the non-negative integer N_χ as follows. If formula (21) admits no solution with an integral divisor σ_χ of degree $p_0 - 1 + t_\chi$ let $N_\chi = 0$. If formula (21) admits solutions with an integral divisor σ_χ of degree $p_0 - 1 + t_\chi$, let

$$N_\chi = i(\sigma_\chi) + t_\chi$$

where $i(\sigma_\chi)$ is the index of σ_χ.

Then $\theta\left[\sum_{j=1}^{s} \beta_j c_j - e_1\right](u;B_1)$ vanishes to order $\sum_{\chi \in R} N_\chi$ at $u = \underline{a}g_0$.

Proof. Let N be the order of vanishing of $\theta\left[\sum_j \beta_j c_j - e_1\right](u;B_1)$ at $\underline{a}g_0$. We show first that $N \geq \sum_{\chi \in R} N_\chi$.

If $N_\chi = 0$ for all χ there is nothing to prove. Assume $N_\chi > 0$ for m characters $\chi^{(1)}, \chi^{(2)}, \ldots, \chi^{(m)}$. For convenience introduce the notation $N^{(k)}, \varepsilon^{(k)}, \beta_j^{(k)}, t^{(k)}$ for $N_{\chi^{(k)}}, \varepsilon_{\chi^{(k)}}, \beta_{\chi^{(k)}j}, t_{\chi^{(k)}}$. Now fix $\chi^{(k)}$ and apply $\underline{a}$ to formula (21). Denote by $\sigma^{(k)}$ an appropriate integral divisor on W_0.

$$\underline{a}g_0 + \underline{a}e_0 - \sum\beta_j^{(k)}\underline{a}(\nu_j^{-1}u_0(x_{0j})) + \underline{a}\varepsilon^{(k)} \equiv \underline{a}u_0(\sigma^{(k)}) + \underline{a}K_0.$$

In this equation replace $\underline{a}(\nu_j^{-1}u_0(x_{0j}))$ by formula (4) and $\underline{a}K_0$ by formula (5). We then obtain

$$\underline{a}g_0 + \underline{a}e_0 - \sum\beta_j^{(k)}\left[u_1(x_{1j}) - \nu_j^{-1}u_1(\underline{a}z_0) - c_j\right] + \underline{a}\varepsilon^{(k)}$$

$$\equiv u_1(\underline{a}\sigma^{(k)}) - (p_0 - 1 + t^{(k)})u_1(\underline{a}z_0)$$

$$+ K_1 + (p_1 - 1)n^{-1}u_1(\underline{a}z_0) + \underline{a}e_0 + e_1.$$

Since $\sum\beta_j^{(k)}\nu_j^{-1} + (p_0 - 1 + t^{(k)}) = n^{-1}(p_1 - 1)$,

$$\sum\beta_j^{(k)}c_j \equiv \sum\beta_j c_j + \sum\chi_j^{(k)}c_j ,$$

and $\sum\chi_j^{(k)}c_j \equiv -\underline{a}\varepsilon^{(k)}$ we obtain

$$(22) \qquad \underline{a}g_0 + \sum\beta_j c_j - e_1 \equiv \sum\beta_j^{(k)}u_1(x_{1j}) + u_1(\underline{a}\sigma^{(k)}) + K_1$$

where the divisor $\sum\beta_j^{(k)}x_{1j} + \underline{a}\sigma^{(k)}$ has degree $\sum\beta_j^{(k)}(n/\nu_j) + n(p_0 - 1 + t^{(k)})$ or $p_1 - 1$.

Suppose first that $m = 1$, that is, there is only one $\chi \in R$ with $N_\chi > 0$. By the Riemann-Roch theorem $N^{(1)}\left(= i(\sigma^{(1)}) + t^{(1)}\right)$ is the number of linearly independent multiples of $-\sigma^{(1)}$ on W_0. Any multiple of $-\sigma^{(1)}$ on W_0 lifts via $\underline{b}$ to a multiple of $-(\underline{a}\sigma^{(1)} + \sum\beta_j^{(1)}x_{1j})$ on W_1.

Since the degree of $(\underline{a}\sigma^{(1)} + \sum \beta_j^{(1)} x_{1j})$ is $p_1 - 1$ the Riemann-Roch theorem implies that the index of this integral divisor is at least $N^{(1)}$ and so $\theta(u;B_1)$ vanishes at $\underline{a}g_0 + \sum \beta_j c_j - e_1$ to order at least $N^{(1)}$; that is, $N \geq N^{(1)}$.

If $m > 1$ formula (22) shows for $k = 2,3, \ldots ,m$ that

$$\sum_j \beta_j^{(k)} u_1(x_{1j}) + u_1(\underline{a}\sigma^{(k)}) \equiv \sum_j \beta_j^{(1)} u_1(x_{1j}) + u_1(\underline{a}\sigma^{(1)}).$$

For $k = 2,3, \ldots ,m$ let $H^{(k)}$ be a function on W_1 whose divisor is as follows.

$$(H^{(k)}) = (\underline{a}\sigma^{(k)} + \sum \beta_j^{(k)} x_{1j}) - (\underline{a}\sigma^{(1)} + \sum \beta_j^{(1)} x_{1j}).$$

Let $H^{(1)}$ be a non-zero constant function. Since the divisor of $H^{(k)}$ is invariant under G, $H^{(k)}$ corresponds to a character which will now be shown to be $\chi^{(k)}/\chi^{(1)}$.

For each k let $\alpha^{(k)}$ be a function which corresponds to $\chi^{(k)}$. We may write the divisor of $\alpha^{(k)}$ as follows.

$$(\alpha^{(k)}) = \underline{a}(D_0^{(k)}) + \sum_j \chi_j^{(k)} x_{1j}.$$

where $D_0^{(k)}$ may include points of X_0. This is possible since $\nu_j x_{1j} = \underline{a}x_{0j}$. By formula (17)

$$u_0(D_0^{(k)}) + \sum_j \chi_j^{(k)} \nu_j^{-1} u_0(x_{0j}) \equiv \varepsilon^{(k)}.$$

Substituting this into formula (21) we have

$$g_0 + e_0 + \sum(\chi_j^{(k)} - \beta_j^{(k)})\nu_j^{-1}u_0(x_{0j}) + u_0(D_0^{(k)}) \equiv u_0(\sigma^{(k)}) + K_0$$

for $k = 1,2, \dots ,m$. Thus for each $k = 2,3, \dots ,m$ there is a function on W_0 whose divisor is

$$\sum_j(\chi_j^{(1)} - \chi_j^{(k)} - \beta_j^{(1)} + \beta_j^{(k)})\nu_j^{-1}x_{0j}$$

$$+D_0^{(1)} - \sigma^{(1)} - D_0^{(k)} + \sigma^{(k)}$$

since $(\chi_j^{(1)} - \chi_j^{(k)}) - (\beta_j^{(1)} - \beta_j^{(k)}) \equiv 0 \pmod{\nu_j}$.

If we lift this function, via $\underline{b}$, to W_1 the lifted function has the divisor

$$\sum_j(\chi_j^{(1)} - \chi_j^{(k)} - \beta_j^{(1)} + \beta_j^{(k)})x_{1j}$$

$$+ \underline{a}(D_0^{(1)} - \sigma^{(1)} - D_0^{(k)} + \sigma^{(k)})$$

which is the divisor of $H^{(k)}\alpha^{(1)}/\alpha^{(k)}$. Thus $H^{(k)}\alpha^{(1)}/\alpha^{(k)}$ corresponds to the identity character, and so $H^{(k)}$ corresponds to the character $\chi^{(k)}/\chi^{(1)}$.

Let $L^{(k)}$, $k = 1,2, \dots ,m$ be the space of multiples of $-\sigma^{(k)}$ on W_0. Then $N^{(k)} = \dim L^{(k)}$ by the Riemann-Roch theorem. Let $\underline{a}L^{(k)}$ be the lifts of the functions in $L^{(k)}$ to W_1 via $\underline{b}$. Thus $\underline{a}L^{(k)}$ is the span of $N^{(k)}$ linearly independent functions which are multiples of $-(\underline{a}\sigma^{(k)} + \sum\beta_j^{(k)}x_{1j})$. Let $H^{(k)}\underline{a}L^{(k)}$ be the family of functions obtained by multiplying the functions in $\underline{a}L^{(k)}$ by $H^{(k)}$. Then every function in $H^{(k)}\underline{a}L^{(k)}$ is a multiple of

$-(\underline{a}\sigma^{(1)} + \sum\beta_j{}^{(1)}x_{1j})$. Thus the families $H^{(1)}\underline{a}L^{(1)}$, $H^{(2)}\underline{a}L^{(2)}, \ldots, H^{(m)}\underline{a}L^{(m)}$ represent $\sum_{k=1}^{m} N^{(k)}$ linearly independent multiples of $-(\underline{a}\sigma^{(1)} + \sum\beta_j{}^{(1)}x_{1j})$ since the $H^{(k)}$'s correspond to different characters. As in the case $m = 1$, we now conclude that $\theta(u;B_1)$ vanishes at $\underline{a}g_0 + \sum\beta_j c_j - e_1$ to order at least $\sum_{k=1}^{m} N^{(k)}$; that is, $N \geq \sum_{k=1}^{m} N^{(k)}$.

The proof that $N \leq \sum_{k=1}^{m} N^{(k)}$ is essentially the reverse of the above argument. Suppose then that $\theta[\sum\beta_j c_j - e_1](u;B_1)$ vanishes to order N at $u = \underline{a}g_0$. If $N = 0$ there is nothing to prove so assume $N > 0$. By the Riemann vanishing theorem there is an integral divisor, ξ_1, of degree $p_1 - 1$ on W_1 so that

$$(23)\quad \underline{a}g_0 + \sum\beta_j c_j - e_1 \equiv u_1(\xi_1) + K_1.$$

By the solution to the Jacobi inversion problem there is an integral divisor of degree p_0 on W_0 so that

$$(24)\quad g_0 + e_0 - \sum\beta_j\nu_j{}^{-1}u_0(x_{0j}) \equiv u(\xi_0) + K_0.$$

Applying $\underline{a}$ to formula (24) yields

$$\underline{a}g_0 + \underline{a}e_0 - \sum\beta_j\underline{a}(\nu_j{}^{-1}u_0(x_{0j})) = \underline{a}u_0(\xi_0) + \underline{a}K_0.$$

Now apply formulas (4) and (5).

$$\underline{a}g_0 + \underline{a}e_0 - \sum\beta_j[u_1(x_{1j}) - \nu_j^{-1}u_1(\underline{a}z_0) - c_j] \equiv$$

$$u_1(\underline{a}\xi_0) - p_0u_1(\underline{a}z_0) + K_1 + (p_1 - 1)n^{-1}u_1(\underline{a}z_0) + \underline{a}e_0 + e_1.$$

Now eliminate $\underline{a}g_0$ between this last equation and formula (23) noting that in the coefficient of $u_1(\underline{a}z_0)$ we have

$$\sum\beta_j\nu_j^{-1} + p_0 - (p_1-1)n^{-1} = 1 - t^{(0)}$$

where $t^{(0)}$ is defined by

$$\sum\beta_j\nu_j^{-1}n = r/2 - t^{(0)}n.$$

We then obtain

$$u_1(\xi_1) \equiv \sum\beta_j u_1(x_{1j}) + u_1(\underline{a}\xi_0) + (t^{(0)} - 1)u_1(\underline{a}z_0).$$

Let

$$(25)\quad D_1 = \sum\beta_j x_{1j} + \underline{a}\xi_0 + (t^{(0)} - 1)\underline{a}z_0.$$

Then D_1 is a divisor of degree $p_1 - 1$, linearly equivalent to ξ_1, and invariant under G. Let f be a multiple of $-D_1$. For each $\chi \in R$ let $f_\chi = \sum_{T\in G}\chi(T^{-1})f \circ T$. Then f_χ is also a multiple of $-D_1$ since each function $f \circ T$ is. Let L_χ be the space of multiples of $-D_1$ which correspond to χ, and let $N_\chi^1 = \dim L_\chi$. Then $N = \sum_{\chi\in R} N_\chi^1$.

If $f(\neq 0)$ is in L_χ then the divisor of f can be written $(f) = \underline{a}\sigma_0 + \sum_{j=1}^{s} f_j x_{1j} - D_1$ where $0 \le f_j < \nu_j$ and σ_0 is an integral divisor on W_0. Since f corresponds to χ we have (for each j)

$$f_j - \beta_j \equiv \chi_j \pmod{\nu_j}$$

or

$$f_j \equiv \chi_j + \beta_j \pmod{\nu_j}.$$

It follows that $f_j = \beta_{\chi j}$. (See formula (14)).
Thus the degree of $\sum f_j x_{1j}$ is $r/2 - t_\chi n$ and so $\underline{a}\sigma_0$ is of degree $n(p_0 - 1 + t_\chi)$. Thus

$$(26) \quad (f) = \underline{a}\sigma_0 + \sum \beta_{\chi j} x_{1j} - D_1$$

where σ_0 is an integral divisor of degree $p_0 - 1 + t_\chi$ on W_0. If f^1 is another function in L_χ then

$$(f^1) = \underline{a}\sigma_0^1 + \sum \beta_{\chi j} x_{1j} - D_1$$

and so $(f^1/f) = \underline{a}(\sigma_0^1 - \sigma_0)$. Since f^1/f corresponds to the identity character we see that $\sigma_0 \equiv \sigma_0^1$ on W_0. It follows that the map $f_0 \to f(f_0 \circ \underline{b})$ is an isomorphism from the space of mulitples of $-\sigma_0$ on W_0 onto L_χ. Consequently $N_\chi^1 = i(\sigma_0) + t_\chi$.

For f in L_χ we combine formulas (24) and (25) to obtain

$$(f) = \underline{a}\sigma_0 + \sum\beta_{\chi j}x_{1j} - \underline{a}(\xi_0 + (t^0 - 1)z_0) - \sum\beta_j x_{1j}.$$

By formula (17) we have

$$\varepsilon_\chi \equiv u_0(\sigma_0) + \sum\beta_{\chi j}\nu_j^{-1}u_0(x_{0j} - u_0(\xi_0) - \sum\beta_j\nu_j^{-1}u_0(x_{0j})$$

or $u(\xi_0) + \sum\beta_j\nu_j^{-1}u_0(x_{0j}) \equiv u_0(\sigma_0) + \sum\beta_{\chi j}\nu_j^{-1}u_0(x_{0j}) - \varepsilon_\chi.$

Substituting this into formula (24) yields

$$g_0 + e_0 - \sum\beta_{\chi j}\nu_j^{-1}u_0(x_{0j}) + \varepsilon_\chi \equiv u(\sigma_0) + K_0.$$

We conclude that if $N_\chi^1 > 0$ then χ satisfies the conditions required in the statement of the theorem for formula (21) to have a solution. Consequently if $N_\chi^1 > 0$ then χ is in the list $\chi^{(1)}, \chi^{(2)}, \ldots, \chi^{(m)}$ given at the beginning of the proof. Consequently $N = \sum_{\chi\in R} N_\chi^1 \leq \sum_{k=1}^{m} N^{(k)}$. q.e.d.

Since $\sum_{\chi\in R} t_\chi = 0$ either all t_χ are zero or else there is a t_χ which is positive. Considering these two possibilities yields several corollaries.

Corollary 1. Suppose $t_\chi = 0$ for all $\chi \in R$. Then there is an exponential function of $u, E(u)$, and a constant $\ell_\beta \neq 0$ so that

(27) $E(u)\theta[\sum\beta_j c_j - e_1](\underline{a}u; B_1)$

$$= \ell_\beta \prod_{\chi \in R} \theta[e_0 - \sum \beta_{\chi j} \nu_j^{-1} u_0(x_{0j}) + \varepsilon_\chi](u;B_0).$$

Moreover, as an nth order theta-function for $J(W_0)$ each side of formula (27) has a (1/2)-integer theta characteristic which is independent of $\beta = (\beta_1, \beta_2, \ldots, \beta_s)$.

Proof: By Lemma 3 there is an exponential function $E(u)$ so that $E(u)\theta[\sum_j \beta_j c_j - e_1](\underline{a}u;B_1)$ is an nth order theta function for $J(W_0)$. By the same generalization of transformation theory mentioned after the statement of Lemma 3, the theta-characteristic of this n^{th} order function is rational since $[\sum_j \beta_j c_j - e_1]$ is a rational theta-characteristic.

We now consider the theta-characteristic of the product on the right hand side of formula (27). Since theta-characteristics add when theta functions are multiplied, the characteristic in question is

$$(28) \quad \sum_{\chi \in R} [e_0 - \sum_j \beta_{\chi j} \nu_j^{-1} u_0(x_{0j}) + \varepsilon_\chi].$$

As in the proof of lemma 7

$$\sum_R \sum_j \beta_{\chi j} \nu_j^{-1} u_0(x_{0j}) = \sum_j n \nu_j^{-1}((\nu_j^2 - \nu_j)/2)\nu_j^{-1} u_0(x_{0j})$$

a sum independent of β. By formula (6) and note (8)

$$ne_0 = \sum (n/2)(\nu_j - 1)\nu_j^{-1} u_0(x_{0j}).$$

Thus

$$\sum_{\chi\in R}[e_0 - \sum_j \beta_{\chi j}\nu_j^{-1}u_0(x_{0j}) + \varepsilon_\chi] = \sum \varepsilon_\chi .$$

Since $\chi \to \varepsilon_\chi$ is a homomorphism and $\prod_{\chi\in R}\chi$ is an element of order two we see that the characteristic in formula (28) is a (1/2)-integer characteristic.

Now fix $g_0 \in J(W_0)$. Since $t_\chi = 0$ for all χ, σ_χ has degree $p_0 - 1$ for all χ such that formula (21) admits a solution. Thus

$$\theta(g_0 + e_0 - \sum \beta_{\chi j}\nu_j^{-1}u_0(x_{0j}) + \varepsilon_\chi ; B_0)$$

vanishes to order N_χ $(= i(\sigma_\chi))$ whenever formula (21) has a solution. Consequently, the product on the right hand side of of formula (27) has order $N = \sum N_\chi$ at g_0.

Since $\theta[\sum \beta_j c_j - e_1](u;B_1)$ has order N at $\underline{a}g_0 \in J(W_1)$, the left hand side of (27) has order at least N considered as a function on $J(W_0)$. Thus the quotient of the left hand side by the product on the right hand side is an entire function on $\mathbb{C}^{p_0}$. Because the theta-characteristics involved are rational, a suitable power of the quotient is an entire function automorphic with respect to the periods defining $J(W_0)$. Thus the quotient is a constant, ℓ_β. That $\ell_\beta \neq 0$ follows from the theorem since the two sides of (27) are non-zero for the same values of g_0, and there are clearly values of g_0 which make the product non-zero. q.e.d.

Corollary 2: If $\underline{b} : W_1 \to W_0$ is an unramified abelian cover, then

$$E(u)\theta[e_1](\underline{a}u;B_1) = \ell \prod_{\varepsilon \in \ker \underline{a}} \theta[\varepsilon](u;B_0)$$

where $2e_1 \equiv 0$.

Proof: This is now immediate since the map $\chi \to \varepsilon_\chi$ is an isomorphism of R onto the kernel of $\underline{a}$. q.e.d.

Corollary 3: Suppose there is a $\chi \in R$ so that $t_\chi \neq 0$. Then $\theta[\sum \beta_j c_j - e_1](u;B_1)$ vanishes to order at least $\sum_{\chi \in R} \max(0,t_\chi)$ for any u in $\underline{a}J(W_0)$. For the general point on $\underline{a}J(W_0)$ this lower bound is achieved.

Proof: For each $\chi \in R$ reconsider formula (21)

$$(21) \quad g_0 + e_0 - \sum \beta_{\chi j} \nu_j^{-1} u_0(x_{0j}) + \varepsilon_\chi \equiv u_0(\sigma_\chi) + K_0$$

where we wish to solve this equation with an integral divisor σ_χ of degree $p_0 - 1 + t_\chi$. If $t_\chi \geq 1$ this is always possible regardless of what g_0 is. Consequently, for any $u = \underline{a}g_0$,

$$N \geq \sum_{t_\chi \geq 1} N_\chi = \sum_{t_\chi \geq 1} (i(\sigma_\chi) + t_\chi) \geq \sum_{\chi \in R} \max(0,t_\chi).$$

To prove the last assertion we show that there are $g_0 \in J(W_0)$ so that $N_\chi = \max(0,t_\chi)$ for all χ. Since $N_\chi = i(\sigma_\chi) + t_\chi$ we need show that there are g_0's which satisfy, for any χ, the following: (i) if $t_\chi \leq 0$ the formula (21) admits no solution σ_χ, and (ii) if $t_\chi > 0$ then $i(\sigma_\chi) = 0$. But for each χ it is easily seen that the set of g_0's not satisfying these requirements lie on a set of co-dimension at least one in $J(W_0)$. For if $t_\chi \leq 0$ then σ_χ is required to have degree $p_0 - 1$ or less. If $t_\chi > 0$ then σ_χ has degree $p_0 - 1 + t_\chi \geq p_0$. Since the general integral divisor of degree $\geq p_0$ has index zero, the proof of the corollary is complete.

q.e.d.

Remarks: A necessary but not sufficient condition that Corollary 3 be applicable to a ramified cover is that $r \geq 2n$. This follows from formula (20).

In case $p_0 = 0$ Corollary 1 says that on certain $(1/2n)$-periods $\theta(u;B_1)$ is non-zero, and Corollary 3 says that on other $(1/2n)$-periods $\theta(u;B_1)$ vanishes to various orders. (See note 4) This generalizes the hyperelliptic situation.

PART II

1. Introduction. In Part II of this paper we will apply the results of Part I to particular cases. The most simple case where $n = 2$ will be considered in some detail in addition to other cases where the surfaces have genus two, three and five. Besides illustrating the general theory, these cases are chosen because it will be shown in Part III that the derived vanishing properties of the theta function sometime characterize those surfaces admitting the particular abelian group.

Before examining the particular cases we must dispose of some preliminary material, some of which is included in this introductory section. In section 2 we show that for a large class of covers, all the vanishing properties of Theorem 1, Part I, can be quickly derived after the case $p_0 = 0$ has been investigated.

The notation of Part I will be continued.

The first of the two preliminary considerations to be taken up in this introduction concerns explicit matrix formulations for the map $\underline{a} : J(W_0) \to J(W_1)$. Suppose canonical homology bases, $\gamma_1, \ldots, \gamma_{2p_0}$, and $\gamma_1', \ldots, \gamma_{2p_1}'$ have been chosen on W_0 and W_1 respectively. The intersection matrices $(\gamma_i \times \gamma_j)^{2p_0 \times 2p_0}$

$= J_0$ and $(\gamma_i' \times \gamma_j')^{2p_1 \times 2p_1} = J_1$ are

$$J_i = \begin{bmatrix} 0 & E \\ -E & 0 \end{bmatrix} \quad i = 0,1$$

where E is the appropriate identity matrix.

We now define a map, α, from singular one-chains on W_0 into singular one-chains on W_1. If γ is a small arc defined in a parametric disc of W_0 and $\underline{b}$ is unramified over this disc, then $\alpha\gamma$ is the n copies of γ in $\underline{b}^{-1}(\gamma)$, each copy oriented the same way as γ. If $\underline{b}$ is branched over γ then the definition is suitably modified. Now extend α to arbitrary one-chains by linearity.

If $H_1(W_i)$ is the first homology group (over the integers) on W_i, $i = 0, 1$, then we can consider α also as a map from $H_1(W_0)$ into $H_1(W_1)$. Thus there is a $2p_0 \times 2p_1$ integer matrix (α_{ij}) so that

$$\alpha\gamma_i = \sum_{j=1}^{2p_1} \alpha_{ij}\gamma_j' \quad i = 1,2, \dots ,2p_0.$$

Now define a map $\underline{a} : \mathbb{C}^{p_0} \to \mathbb{C}^{p_1}$ as follows. If γ is a singular one-chain on W_0 and $du_i = (du_{i1}, du_{i2}, \dots ,du_{ip_i})^{\sim}$, $i = 0,1$ is a basis of analytic differentials dual to the given canonical homology bases, let

$$\underline{a} \int_\gamma du_0 = \int_{\alpha\gamma} du_1.$$

Since α is a mapping of homology the map $\underline{a} : \mathbb{C}^{p_0} \to \mathbb{C}^{p_1}$ takes periods into periods and so reduces to the original $\underline{a} : J(W_0) \to J(W_1)$ by reducing modulo periods. For $j = 0,1$ let

$$\Omega_j = (\pi i E, B_j)^{p_j \times 2p_j}$$

and let $\Omega_{0i} = \int_{\gamma_i} du_0$ and $\Omega_{1j} = \int_{\gamma_j'} du_1$.

Since

$$\underline{a} \int_{\gamma_i} du_0 = \int_{\sum_j \alpha_{ij}\gamma_j'} du_1$$

or

$$\underline{a}\Omega_{0i} = \sum_{j=1}^{2p_1} \Omega_{1j}\alpha_{ij}$$

we see that

$$\underline{a}\Omega_0 = \Omega_1 \tilde{\alpha} \tag{1}$$

where in this context $\underline{a}$ is a $p_1 \times p_0$ complex matrix and $\tilde{\alpha}$ is a $2p_1 \times 2p_0$ integer matrix.

Since the intersection matrix of the $\alpha\gamma_i$'s is seen to be nJ_0 it follows that

$$nJ_0 = (\alpha\gamma_i \times \alpha\gamma_j) = \alpha J_1 \tilde{\alpha}. \tag{2}$$

Let

$$\alpha = \begin{bmatrix} \alpha_1 & \alpha_2 \\ \alpha_3 & \alpha_4 \end{bmatrix}$$

where each α_j is a $p_0 \times p_1$ matrix. Then formula (2) leads to the equations

(3)
$$\begin{cases} \alpha_2\tilde{\alpha}_1 - \alpha_1\tilde{\alpha}_2 = 0 \\ \alpha_4\tilde{\alpha}_1 - \alpha_3\tilde{\alpha}_2 = nE \\ \alpha_4\tilde{\alpha}_3 - \alpha_3\tilde{\alpha}_4 = 0 . \end{cases}$$

With these considerations we will indicate how Lemma 3 of Part I is proven.

<u>Lemma 1:</u> Let

(1)
$$\underline{a}\Omega_0 = \Omega_1\tilde{\alpha}$$

describe the map $\underline{a} : J(W_0) \to J(W_1)$ where

(2)
$$nJ_0 = \alpha J_1\tilde{\alpha}.$$

Then

$$\theta\begin{bmatrix} g \\ h \end{bmatrix}(\underline{a}u; B_1) = \ell \text{ Exp } \{\pi i \tilde{u} \frac{-\tilde{a}\tilde{\alpha}_2}{(\pi i)^2} u\} \; \phi\;(u)$$

where $\phi(u)$ is an nth order theta function for $J(W_0)$ with characteristic $\begin{bmatrix}\hat{g}\\ \hat{h}\end{bmatrix}$ where[1)]

$$(4) \qquad \begin{pmatrix}\tilde{\hat{g}}\\ -\tilde{\hat{h}}\end{pmatrix} = \alpha \begin{pmatrix}\tilde{g}\\ -\tilde{h}\end{pmatrix} - \frac{1}{2}\begin{pmatrix}\mathrm{Sp}\ \alpha_1\ \tilde{\alpha}_2\\ \mathrm{Sp}\ \alpha_3\ \tilde{\alpha}_4\end{pmatrix}$$

Proof: If $n = 1$ the result is simply a restatement of the transformation theory for first order theta functions under first order transformations. If we consider formulas (1) and (2) formally, forgetting for the moment the present context, and assume that $\underline{a}$ and α are square matrices (i.e., $p_0 = p_1$), then the result is simply a restatement of the transformation theory for first order theta functions under n^{th} order transformations[2)].

The slight degree of generalization needed for the proof of this lemma follows from the observation that the classical proofs go through when $p_0 \neq p_1$; that is, all the matrix computations work when $\underline{a}$ and α are rectangular and formulas (1) and (2) are satisfied. We omit the details.

q.e.d.

Another classical technique which will be used is that of varying the conformal structure of a Riemann surface so that curves on the varying surfaces are "squeezed to points" and the

---

1) If A is a square matrix, $\mathrm{Sp}A$ will be the column vector made up of the diagonal elements of A taken in the same order.

2) See Krazer [15] p. 166.

limit "surface" is the union of two or more punctured surfaces. We will formulate the procedure a little more precisely. Other formulations of this procedure may be found in [13] and [17].

In the annulus $A' = \{\rho^{-1} < |z| < 1\}$ let $\mu (= \mu(z)\frac{d\bar{z}}{dz})$ be a radially symmetric Beltrami differential which defines a conformal structure on A' so that in this new structure A' is conformally equivalent to $\{0 < M < |z| < \infty\}$. Necessarily $|\mu(z)| \to 1$ as $|z| \to 1$. In the annulus $A = \{\rho^{-1} < |z| < \rho\}$ extend the definition of μ by reflection in the unit circle. For $0 \le t < 1$ let $\mu_t = t\mu$ and let $A(t)$ be the annulus with a conformal structure defined by μ_t. Then $A(t)$ $(0 \le t < 1)$ is conformally equivalent to another annulus. Now extend the definition of μ_t to the Riemann sphere by letting μ be zero outside A, and let $S(t)$ be the Riemann sphere in the conformal structure defined by μ_t. Then $S(t)$ is conformally a sphere, but as $t \to 1$, $S(t)$ "approaches" two punctured spheres. In this case we will say the unit circle is "squeezed to a point" as $t \to 1$.

Now let W be a Riemann surface of genus p. Let $C_1, C_2, \ldots, C_r$ be r simple closed analytic curves each of which divides W into two components. Suppose there are pairwise disjoint annuli, A_j, $j = 1, 2, \ldots, r$ so that A_j is conformally equivalent to $\{\rho_j^{-1} < |z| < \rho_j\}$ and C_j corresponds to the unit circle in A_j. We obtain a global Beltrami differential μ by setting $\mu = 0$ in $W - \bigcup_{j=0}^{r} A_j$ and μ equal in each A_j to a differential like the one in the previous paragraph. For each t, $0 \le t < 1$ $t\mu$ defines a conformal structure on W, and we denote the corresponding Riemann surface $W(t)$. As $t \to 1$ the curves

C_j are squeezed to points and $W(t) \to W(1)$ where $W(1)$ is a union of $r + 1$ punctured surfaces.

Suppose that $A_{01}, \dots, A_{0p_0}, A_{11}, \dots, A_{1p_1}, \dots$ $A_{r1}, \dots, A_{rp_r}, B_{01}, \dots, B_{0p_0}, \dots, B_{r1}, \dots, B_{rp_r}$ is a canonical homology basis of W so that $A_{j1}, \dots, A_{jp_j}$, $B_{j1}, \dots, B_{jp_j}$ is a canonical basis modulo dividing cycles for the jth component of $W - \bigcup_{i=1}^{r} A_i$, $j = 0,1,\dots,r$. Let $(du_0(t)^{\sim}, du_1(t)^{\sim}, \dots, du_r(t)^{\sim})^{\sim}$ be the basis of analytic differentials of $W(t)$ duel to the given canonical homology basis where

$$du_j(t) = \begin{pmatrix} du_{j1}(t) \\ du_{j2}(t) \\ \cdot \\ \cdot \\ \cdot \\ du_{jp_j}(t) \end{pmatrix}.$$

Let the corresponding period matrix for $W(t)$ be $(\pi i E, B(t).)^{p \times 2p}$. If $k \neq j$ then

$$\int_{A_{k\ell}} du_j(t) = 0 \quad \text{and} \quad \int_{B_{k\ell}} du_j(t) \to 0$$

as $t \to 1$. Thus as $t \to 1$

$$B(t) \to \text{diag}\,(B_{p_0}, B_{p_1}, \dots, B_{p_r})$$

where $(\pi i E, B_{p_j})^{p_j \times 2p_j}$ is the period matrix for the jth component of $W(1)$. In the kth component of $W(0) - \bigcup_{i=1}^{r} A_i$

the conformal structure of $W(t)$ is unchanged as t varies. Consequently, if γ is a path lying in the kth component of $W(0) - \bigcup_{i=1}^{r} A_i$ and $j \neq k$ then $\int_\gamma du_j(t) \to 0$ as $t \to 1$. Finally, if

$$\theta\begin{bmatrix} g_0 & g_1 & \cdots & g_r \\ h_0 & h_1 & \cdots & h_r \end{bmatrix}(u;B(t)) \quad \text{is a first order}$$

theta function for $W(t)$ where $\begin{bmatrix} g_i \\ h_i \end{bmatrix}$ is a p_i-theta characteristic, then as $t \to 1$

$$\theta\begin{bmatrix} g_0 & g_1 & \cdots & g_r \\ h_0 & h_1 & \cdots & h_r \end{bmatrix}(0;B(t)) \to \prod_{j=0}^{r} \theta\begin{bmatrix} g_j \\ h_j \end{bmatrix}(0;B_{p_j})$$

where $\theta\begin{bmatrix} g_j \\ h_j \end{bmatrix}(u;B_{p_j})$ is a first order theta function for the jth component of $W(1)$.

2. Completely Ramified Abelian Covers. Let $\underline{b} : W_1 \to W_0$ be an arbitrary abelian cover. If the corresponding function fields are M_0 and M_1, $M_0 \subset M_1$, let M_{UA} be the maximum unramified abelian extension of M_0 in M_1. Thus $M_0 \subset M_{UA} \subset M_1$. Then $\underline{b} : W_1 \to W_0$ admits a corresponding factorization $W_1 \to W_{UA} \to W_0$. If $W_{UA} = W_1$ then, of course, the original cover was unramified. If $W_{UA} = W_0$, the covering $\underline{b}$ will be called completely ramified. In any case where $M_{UA} \neq M_1$ the cover $W_1 \to W_{UA}$ is completely ramified.

Suppose that $\underline{b} : W_1 \to W_0$ is a completely ramified abelian cover. Suppose W_0 has ramification occurring at $x_{01}, x_{02}, \ldots, x_{0s}$, $s \geq 1$. Then the covering $W_1 \to W_0$ is determined by a representation μ of the fundamental group of $W - \{x_{01}, \ldots, x_{0s}\}$ into G, the Galois group of the cover.[3] Let a_j be the element of G which μ assigns to a path which "circles" x_{0j}, $j = 1,2, \ldots ,s$.[4] Then we assert that G is

---

3) We will describe μ a little more carefully. Let y_0 be a point of $W_0 - \{x_{01}, x_{02}, \ldots, x_{0s}\}$ which serves as the base point for the fundamental group. Then if we number the n points of $\underline{b}^{-1}(y_0)$, say $\{y_1, y_2, \ldots, y_n\}$, every closed path γ based at y_0 induces a permutation, $\mu(\gamma)$, of $\underline{b}^{-1}(y_0)$. The image of μ is a regular transitive subgroup of S_n, the symmetric group on n objects, and this subgroup is isomorphic to G. The cover transformations also induce a regular transitive representation of G in S_n. Since G is abelian these two subgroups of S_n are identical. We identify this subgroup with G for the purposes of this discussion.

4) Since G is abelian all conjugates of a_j equal a_j and so there is no ambiguity in the choice of a_j.

generated by the elements $a_1, a_2, \ldots, a_s$. For let H be the subgroup of G generated by these elements. If H is not G then the cover $W_1 \to W_0$ $(= W_1/G)$ admits a proper factorization $W_1 \to W_1/H \to W_0$. Since the cover $W_1/H \to W_0$ is seen to be unramified, we have reached the desired contradiction.

The following lemma gives a topological description of completely ramified covers.

Lemma 2: Let $\underline{b} : W_1 \to W_0$ be a completely ramified abelian cover. Then there is a disc $\Delta_0 \subset W_0$ (let $\Delta_1 = \underline{b}^{-1}(\Delta_0)$) with the following properties. All the ramification of $\underline{b}$ occurs over Δ_0 and $W_1 - \Delta_1$ divides into n components each of which is homeomorphic to $W_0 - \Delta_0$ under $\underline{b}$.

Proof: If $p_0 = 0$ there is nothing to prove so assume $p_0 \geq 1$. The cover $\underline{b} : W_1 \to W_0$ is described by a representation μ from the fundamental group of $W_0 - \{x_{01}, \ldots, x_{0s}\}$ into G. Let Δ_0 be any disc containing $x_{01}, \ldots, x_{0s}$. Let $\gamma_1, \ldots, \gamma_{2p_0}$ be simple closed curves on W_0 whose homotopy classes generate the fundamental group of W_0 and supposed that no curve, γ_ℓ, intersects Δ_0. Again let a_j, $j = 1, \ldots, s$, be the elements of G which corresponds to x_{0j} under μ. Suppose the element which corresponds to γ_ℓ under μ is

$$a_1^{\alpha_{\ell 1}} a_2^{\alpha_{\ell 2}} a_3^{\alpha_{\ell 3}} \ldots a_s^{\alpha_{\ell s}}$$

where $\alpha_{\ell j} \geq 0$. Let $N = \sum_{\ell, j} \alpha_{\ell j}$. To prove the lemma by induction on N it suffices to show that by modifying γ_ℓ within its homotopy class on W_0 the number N can be decreased by one. For if

$N = 0$ a disc Δ_0 corresponding to that choice of the γ_ℓ's will satisfy the conclusion of the lemma.

To show that N can be decreased by one cut W_0 along γ_ℓ where $\alpha_{\ell j} \neq 0$. The resulting cut surface has two boundary contours corresponding to the two sides of the non-dividing cycle γ_ℓ. If τ is a little circle around x_{0j} which includes no other x_{0k}, we can find a path σ from a point of τ to one of the two contours of $W_0 - \gamma_\ell$ so that the path $\gamma_\ell \sigma \tau^{-1} \sigma^{-1}$ is homotopic on $W_0 - X_0$ to a simple closed curve γ_ℓ'. γ_ℓ' is clearly homotopic to γ_ℓ on X_0. Now pick a Δ_0 complementary to the curves $\gamma_1, \ldots, \gamma_{\ell-1}, \gamma_\ell', \gamma_{\ell+1}, \ldots, \gamma_{2p_0}$. For this new Δ_0 N is reduced by one.

q.e.d.

We now turn to a characterization of completely ramified abelian covers, among arbitrary abelian covers, in terms of the map $\underline{a} : J(W_0) \to J(W_1)$.

<u>Lemma 3</u>: Suppose $\underline{b} : W_1 \to W_0$ is an arbitrary abelian cover. $\underline{b}$ is completely ramified if and only if $\underline{a} : J(W_0) \to J(W_1)$ is an isomorphism.

<u>Proof</u>: We show that M_0 admits a non-trivial unramified abelian extension if and only if $\ker \underline{a} \neq (0)$.

Suppose $\ker \underline{a} \neq (0)$. Then there is a divisor D_0 on W_0 of degree zero which is not principal so that $\underline{a}u_0(D_0) = u_1(\underline{a}D_0)$ is principal. If $(f_1) = \underline{a}D_0$ then (f_1) is invariant under G and so $f_1^n (= g_0)$ is in M_0. Consequently nD_0 is principal on W_0 and so $\sqrt[n]{g_0}$ $(\in M_1)$ generates an unramified extension of M_0.

The converse is obtained by reversing the above argument.

q.e.d.

At this point we can complete the proof of Lemma 2, Part I when M_1 is an abelian extension of M_0. Considering the covers $W_1 \to W_2 \to W_0$ it is clear from the definition of $\underline{a}$ that if we have maps $\underline{a}_{ij} : J(W_i) \to J(W_j)$ then $\underline{a}_{21} \circ \underline{a}_{02} = \underline{a}_{10}$. If $W_2 = W_{UA}$ then $\underline{a}_{21}$ is an isomorphism, and so $\ker \underline{a}_{10} = \ker \underline{a}_{02}$. Since we know this lemma to be true for unramified abelian covers (Part I, Section 5), the proof in the general abelian case follows. (In fact, the above arguments show the lemma true for arbitrary covers.)

We now use the information of Lemma 2 to obtain information about the points c_j in $J(W_1)$ (Part I, Section 4.)

<u>Lemma 4</u>: Let $\underline{b} : W_1 \to W_0$ be a completely ramified abelian cover and suppose we choose Δ_0 as in Lemma 2. Let $\nu_j^{-1} u_0(x_{0j})$ (Part I, Section 4) equal

$$(1/\nu_j) \int_{z_0}^{x_{0j}} du_0$$

where the path of integration lies within Δ_0. Let $n^{-1} u_1(\underline{a}z_0)$ (Part I, Section 4) equal

$$(1/n) \int_{nz_1}^{\underline{a}z_0} du_1$$

where the paths are fixed and contained within Δ_1.

Then for all $\chi \in R$ we have $\sum \chi_j c_j = 0$ and consequently, $\varepsilon_\chi = 0$ for all χ.

Proof: Let $p_1 = np_0 + q$ where q is the genus of Δ_1. If $A_1^0, \ldots, A_{p_0}^0, B_1^0, \ldots, B_{p_0}^0$ is a canonical homology basis for W_0 (du_0 the p_0-vector of analytic differentials dual to this bases) let $A_1^1, \ldots, A_{p_0}^1, A_1^2, \ldots, A_{p_0}^2, \ldots, A_1^n, \ldots, A_{p_0}^n, B_1^1, \ldots, B_{p_0}^n$ be the lifts of $A_1^0, \ldots, B_{p_0}^0$ to the n components of $W_1 - \Delta_1$. Let $\bar{A}_1, \ldots, \bar{A}_q, \bar{B}_1, \ldots, \bar{B}_q$ be a canonical basis of Δ_0. Then $\bar{A}_1, \ldots, \bar{A}_q, A_1^1, \ldots, A_{p_0}^n, \bar{B}_1, \ldots, B_{p_0}^n$ is a canonical homology basis for W_1. Denote the dual basis by du_1 where $du_1 = (\tilde{du}_q, \tilde{du}_{11}, \tilde{du}_{12}, \ldots, \tilde{du}_{1n})^\sim$ where du_q is a q-vector and $du_{1\ell}$ is a p_0 vector of differentials.

With respect to these canonical homology bases on W_0 and W_1 it is clear that $\alpha^{2p_0 \times 2p_1}$ has the form

$$\begin{bmatrix} \alpha_1 & 0 \\ 0 & \alpha_4 \end{bmatrix}$$

where $\alpha_1 = \alpha_4 = [0, E, E, \ldots, E]$ where 0 is the $p_0 \times q$ zero matrix and E is the $p_0 \times p_0$ identity matrix. Consequently if $e \in J(W_0)$ has period characteristic $\begin{pmatrix} g \\ h \end{pmatrix}$ then $\underline{a}e$ has period characteristic

$$\begin{pmatrix} 0 & g & g & \ldots & g \\ 0 & h & h & \ldots & h \end{pmatrix}. \tag{5}$$

We now show that for any j, $c_j(\varepsilon J(W_1))$ has a period characteristic of the form

$$\begin{pmatrix} g_1 & 0 & 0 & \ldots & 0 \\ h_1 & 0 & 0 & \ldots & 0 \end{pmatrix}. \tag{6}$$

By formula (4) of Part I

$$(7) \qquad c_j \equiv u_1(x_{1j}) - \underline{a}(\nu_j^{-1}u_0(x_{0j})) - \nu_j^{-1}u_1(\underline{a}z_0).$$

Now squeeze $\partial\Delta_0$ and the n-components of $\partial\Delta_1$ to $n + 1$ points simultaneously. During the process the period characteristic of c_j is constant since it is rational and a continuous function of the varying surface. In the limit B_1 becomes

$$\text{diag}\ (B_q, B_0', B_0', \ldots, B_0')$$

where B_q is the B matrix for Δ_1 and B_0' is the B matrix for $W_0 - \Delta_0$ in the limit. Moreover, $du_{0\ell} \to 0$ in Δ_1 and $du_0 \to 0$ in Δ_0. Thus in the limit the right hand side of formula (7) becomes a point in $J(W_1)$ of the form $(\tilde{e}_q, 0, 0, \ldots, 0)^{\sim}$. Thus c_j has the indicated period characteristic.

Now consider Lemma 8, Part I. For any χ, $\underline{a}\varepsilon_\chi \equiv -\sum_{j=1}^{s} \chi_j c_j$. Since the left hand side has period characteristic as in formula (5) and the right hand side has period characteristic as in formula (6) both sides are zero. Since $\underline{a}$ is an isomorphism, the proof is complete.

Lemma 5: Continue the hypotheses of Lemma 4. Suppose further that there is a $\beta \in V$, $\beta = (\beta_1, \beta_2, \ldots, \beta_s)$ so that (i) $\sum_{j=1}^{s} n\nu_j^{-1}\beta_j \equiv 0 \pmod n$ and (ii) $\sum \beta_j c_j \equiv 0$ in $J(W_1)$. Then there is a $\chi \in R$ so that $\chi_j = \beta_j$ for all j.

Proof: Suppose first that $p_0 = 0$. By formula (4) of Part I

$$\sum \beta_j u_1(x_{1j}) \equiv \sum \beta_j \nu_j^{-1} u_1(\underline{a}z_0)$$

or

$$\sum \beta_j x_{1j} - (\sum \beta_j \nu_j^{-1})\underline{a}z_0 \equiv 0.$$

Let α be a function on W_1 with the above invariant divisor as its divisor. Then α corresponds to some character χ and by definition $\chi_j = \beta_j$ for all j.

In the general completely ramified case the period characteristic of c_j is uneffected by squeezing $\partial\Delta_0$ and $\partial\Delta_1$ to points. The same is true of the s-tuple $(\chi_1, \ldots, \chi_s)$ since it depends on the topological nature of the covering $\underline{b} : \Delta_1 \to \Delta_0$. Thus the proof for $p_0 = 0$ suffices for the general case.

q.e.d.

It can be shown that hypothesis (i) is necessary.

Lemma 6: Continue the hypotheses of Lemma 4. Suppose further that $\sum_{j=1}^{s} (n\nu_j^{-1}) \equiv 0 \pmod{n}$. Then $2e_1 \equiv 0$ if and only if there exists a χ so that $\chi_j = 1$ for all j.

Proof: This follows immediately from the previous two lemmas and the fact that $\sum c_j = -2e_1$.

Lemma 7: Continue the hypotheses of Lemma 4. Then the (1/2n)-period e_1 of formula (5) Part I has period characteristic of the form

$$\begin{pmatrix} g_1 & 0 & 0 & \ldots & 0 \\ h_1 & 0 & 0 & \ldots & 0 \end{pmatrix}.$$

<u>Proof</u>: We repeat formula (5) of Part I

$$(8) \qquad K_1 \equiv \underline{a}K_0 - (p_1 - 1)n^{-1}u_1(\underline{a}z_0) - \underline{a}e_0 - e_1.$$

On $W_0 - \Delta_0$ consider dividing Jordan curves δ_ℓ, $\ell = 2,3, \ldots, p_0$ which divide $W_0 - \Delta_0$ into p_0 components, each of genus one and each component containing a handle pair of the given canonical homology basis for W_0. Now lift each δ_ℓ to n copies on $W_1 - \Delta_1$, one for each component. Now squeeze $\partial\Delta_0$ and the δ_ℓ's to points on W_0 and squeeze the np_0 lifts of these curves on W_1 simultaneously. In the limit W_0 becomes $p_0 + 1$ punctured surfaces, one of genus zero and the remaining components of genus one. W_1 becomes one surface of genus q (lying over the genus zero surface below) and np_0 surfaces of genus one. Since the vector of Riemann constants for a torus has period characteristic $\binom{1}{1}$, in the limit K_0 has period characteristic

$$\begin{pmatrix} 1 & 1 & 1 & \ldots & 1 \\ 1 & 1 & 1 & \ldots & 1 \end{pmatrix} = \begin{pmatrix} k_1 \\ k_2 \end{pmatrix}$$

As before $n^{-1}u_1(\underline{a}z_0)$ becomes a point in $J(W_1)$ of the form $(\tilde{e},0,0, \ldots ,0)^{\sim}$ where e is a q-vector. In the limit $e_0 = 0$. K_1 becomes

$$\begin{pmatrix} \kappa_1 & k_1 & k_1 & \ldots & k_1 \\ \kappa_2 & k_2 & k_2 & \ldots & k_2 \end{pmatrix}$$

where $\begin{pmatrix} k_1 \\ k_2 \end{pmatrix}$ comes from each set of p_0 tori arising from a component of $W_1 - \Delta_1$. By putting this information into formula (8) the lemma is proved.

q.e.d.

Now let u be a point of $J(W_0)$ with a fixed period characteristic $\binom{g}{h}$; that is;

$$u = \pi i\tilde{h} + B_0\tilde{g}$$

As we squeeze $\partial\Delta_0$ to a points u is a continuous function of the surface. (The δ's of the previous proof will no longer be considered).

$$\theta[\sum \beta_j c_j - e_1](\underline{a}u;B_1)$$

then approaches

$$\theta[\sum \beta_j c_j{}' - e_1{}'](0;B_q)\{\theta[{}^0_0](u;B_0{}')\}^n$$

since $B_1 \to \text{diag}\,(B_q', B_0', B_0', \ldots ,B_0')$. The q-characteristic $[\sum \beta_j c_j{}' - e_1{}']$ is obtained from the p_1-characteristic $[\sum \beta_j c_j - e_1]$ by deleting the last np_0 columns, all of which are zero.

We now summarize this discussion in a theorem.

<u>Theorem 1</u>: Let $\underline{b} : W_1 \to W_0$ be a completely ramified abelian cover. Continue the other hypotheses of Lemma 4. Then the vanish-

ing properties of $\theta[\sum \beta_j c_j - e_1](u;B_1)$ on a general point of $\underline{a}J(W_0)$ are precisely the vanishing properties of $\theta[\sum \beta_j c_j' - e_1'](u;B_q)$ at $u = 0$.

3. <u>Two-Sheeted Covers</u>. Let $\underline{b} : W_1 \to W_0$ be a two-sheeted possibly ramified cover. In this situation W_1 will be called a <u>p_0-hyperelliptic Riemann surface</u> and M_1 will be called a <u>p_0-hyperelliptic function field</u>. If $p_0 = 0$ we are in the hyperelliptic case and if $p_0 = 1$ W_1 is called <u>elliptic-hyperelliptic</u>. If r is the total ramification then

$$p_1 = 2p_0 + (r/2) - 1.$$

In this case $s = r$ and $X_1 = x_{11} + \ldots + x_{1s}$ is a divisor of degree s. Also $2q + 2 = r$.

Choose Δ_0 and Δ_1 in accordance with Lemma 4, and let $z_1 = x_{11}$ as in the classical case. Fix a canonical homology basis on W_0 and choose that part of the canonical homology basis for W_1 lying in Δ_1 exactly as in the classical hyperelliptic case.[5] If the period characteristic of c_j, $j = 1,2, \ldots ,2q + 2$ is denoted (a_{j-1}) and the theta characteristic $[e_1]$ is denoted $[n]$ then squeezing $\partial\Delta_0$ and $\partial\Delta_1$ to 3 points simultaneously shows that

$$[n] = \begin{bmatrix} 1 & 1 & 1 & \ldots & 1 & 0 & \ldots & 0 & 0 & \ldots & 0 \\ 1 & 2 & 3 & \ldots & q & 0 & \ldots & 0 & 0 & \ldots & 0 \end{bmatrix}_2$$ [6]

and the (a_j)'s $j = 1,2, \ldots ,2q + 1$ are analogous in the same way to those given in Krazer [15] p.448. Since $G = Z_2$ $R = \{id, \chi\}$

5) See Krazer [15] p. 445.

6) A subscript 2 on a period or theta characteristic will indicate a (1/2)-integer characteristic.

where the corresponding elements in $V = Z_2 \times Z_2 \times \ldots \times Z_2$ (s times) are $(0,0,0, \ldots ,0)$ and $(1,1,1, \ldots ,1)$.
Let $(\sum^{\sigma} a)$, $\sigma = 1,2, \ldots ,2q + 1$ stand for an arbitrary sum of σ distinct period characteristics, (a_j), none of which is $(a_0) = (0)$.
By Lemma 6 $(\sum^{2q+1} a) = (0)$.

We could use Corollary 3 to Theorem 1 of Part I to derive the usual vanishing properties for the hyperelliptic theta function, but we will confine ourselves to tabulating the results by Theorem 1 of Part II, assuming the classical results. (Krazer [15] p. 459.)

Corollary 1: If W_1 is a p_0-hyperelliptic Riemann surface and $p_1 = 2p_0 + q$, the following table gives the vanishing properties that follow from Corollary 3 of Part I.

If σ equals ($\sigma \leq q$)	then $\theta[n + \sum^{\sigma} a](u;B_1)$ vanishes on $\underline{a}J(W_0)$, in general, to order
q	0
$q - 1$ or $q - 2$	1
$q - 3$ or $q - 4$	2
$q - 5$ or $q - 6$	3
.	.
.	.
.	.

We now use the table to derive the highest order vanishing properties of a p_0-hyperelliptic theta function. As in the hyperelliptic case we distinguish between p_1 (and q) being odd or even.

<u>Corollary 2</u>: If W_1 is a p_0-hyperelliptic Riemann surface where p_1 is odd then $\theta[n](u;B_1)$ vanishes to order $(q + 1)/2$ at a general point of $\underline{a}J(W_0)$. Thus $\theta[n](u;B_1)$ vanishes to order at least $(p_1 + 1)/2 - p_0$ on the 4^{p_0} half-periods of $\underline{a}J(W_0)$.

<u>Corollary 3</u>: If W_1 is a p_0-hyperelliptic Riemann surface where p_1 is even, then $\theta[n](u;B_1)$ and the $2q + 1$ functions $\theta[n + \sum^{1} a](u;B_1)$ vanish at a general point of $\underline{a}J(W_0)$ to order $q/2$. Thus these $2q + 2$ functions vanish to order at least $p_1/2 - p_0$ at the 4^{p_0} half-periods of $\underline{a}J(W_0)$.

We now consider the elliptic-hyperelliptic case of Corollary 2, for $p_1 \geq 5$. Putting in the four half-periods of $\underline{a}J(W_0)$ for u in $\theta[n](\underline{a}u;B_1)$ gives the following.

<u>Corollary 4</u>: Let W_1 be an elliptic-hyperelliptic surface of odd genus, five or more. Then there are four half-integer theta characteristics $[\eta_k]$, $k = 1,2,3,4$ so that $\eta_1 + \eta_2 + \eta_3 = \eta_4$ and $\theta[\eta_k](u;B_1)$ vanishes to order precisely $(p_1 - 1)/2$ at $u = 0$.

<u>Proof</u>: The word "precisely" needs explanation. In the proof of Corollary 3 to Theorem 1, Part I, $\theta[\sum\beta_j c_j - e_1](u;B_1)$ vanishes at $\underline{a}g_0$ to order $\sum N_\chi$ where $N_\chi = t_\chi + i(\sigma_\chi)$ and σ_χ is an integral divisor on W_0. But if $p_0 = 1$ then $i(\sigma_\chi) = 0$ for all integral divisors. Thus $\theta[\sum\beta_j c_j - e_1](u;B_1)$ vanishes to order precisely $\sum_{t_\chi > 0} t_\chi$ when $p_0 = 1$.

q.e.d.

Note that Corollary 4 applied to the case $p_1 = 5$ gives the correct count on the codimension of the elliptic-hyperelliptic locus in Teichmüller space for genus 5, the codimension being 4.

The theta characteristics of the last corollary are of the form

$$\begin{bmatrix} 1 & 1 & 1 & \ldots & 1 & \varepsilon & \varepsilon \\ 1 & 2 & 3 & \ldots & p_1 - 2 & \varepsilon' & \varepsilon' \end{bmatrix}_2$$

where $\begin{bmatrix} \varepsilon \\ \varepsilon' \end{bmatrix}_2$ is an arbitrary half-integer characteristic for genus one. If $p_1 = 3$ there are no special vanishing properties of $\begin{bmatrix} 1/2 & \varepsilon/2 & \varepsilon/2 \\ 1/2 & \varepsilon'/2 & \varepsilon'/2 \end{bmatrix}(u;B_1)$ at $u = 0$ since the theta characteristic is odd. However, the fact that $\theta\begin{bmatrix} 1/2 & 0 & 0 \\ 1/2 & 0 & 0 \end{bmatrix}(\underline{a}\underline{u};B_1)$ vanishes for all $u \in J(W_0)$ and the period characteristic of $\underline{a}\underline{u}$ is of the form $\begin{pmatrix} 0 & g & g \\ 0 & h & h \end{pmatrix}$ shows that $\theta\begin{bmatrix} 1/2 & \varepsilon & \varepsilon \\ 1/2 & \varepsilon' & \varepsilon' \end{bmatrix}(0;B_1)$ vanishes for any characteristic $\begin{bmatrix} \varepsilon \\ \varepsilon' \end{bmatrix}$. Letting f_1 and f_2 be the quarter-periods whose period characteristics are $\begin{pmatrix} 1/2 & 0 & 0 \\ 1/2 & 1/4 & 1/4 \end{pmatrix}$ and $\begin{pmatrix} 1/2 & 1/2 & 1/2 \\ 1/2 & 1/4 & 1/4 \end{pmatrix}$ gives the following corollary

<u>Corollary 5</u>: Let W_1 be an elliptic-hyperelliptic Riemann surface of genus three. Then there are two quarter-periods, f_1 and f_2,[7)] so that i) $f_1 \not\equiv \pm f_2$, ii) $2f_1 \equiv 2f_2 \not\equiv 0$,

---

7) For the pair f_1 and f_2 we could equally well have taken $\begin{pmatrix} 1/2 & 1/4 & 1/4 \\ 1/2 & 0 & 0 \end{pmatrix}$ and $\begin{pmatrix} 1/2 & 1/4 & 1/4 \\ 1/2 & 1/2 & 1/2 \end{pmatrix}$ or $\begin{pmatrix} 1/2 & 1/4 & 1/4 \\ 1/2 & 1/4 & 1/4 \end{pmatrix}$ and $\begin{pmatrix} 1/2 & 3/4 & 3/4 \\ 1/2 & 1/4 & 1/4 \end{pmatrix}$.

iii) $|(2f_1),(f_1+f_2)| = 1$[8] and iv) $\theta(f_1;B_1) = \theta(f_2;B_1) = 0$.

Now we apply Corollary 1 of Theorem 1, Part I, to the p_0-hyperelliptic case where $r \geq 2$. Since $\left(\sum_{j=1}^{2q+1} a\right) = (0)$ we need only consider sums of q (a_j)'s. Select $q+1$ points of X_0 so that the first one is always x_{01} and the others have indexes x_{0j_k}, $k = 1, \ldots, q$. Let $(\sum^q a) = (\sum_{k=1}^{q} a_{(j_k-1)})$. In applying formula (27), Part I, observe that $4e_0 = \sum_1^s u_0(x_{0j})$. Let β be that s-tuple in V with ones in the first and $j_1, j_2, \ldots, j_q$ positions, and let β modified by χ (call it β_χ) be the s-tuple with ones in the other positions. Finally let

$$\sigma = \sum_1^s \beta_j u_0(x_{0j}) - \sum_1^s \beta_{\chi j} u_0(x_{0j}).$$

Formula (27), Part I, then reads

$$\theta[n + \sum^q a](\underline{a}u;B_1) = \tag{9}$$

$$\ell_\beta \theta[(1/4)\sigma](u;B_0)\theta[-(1/4)\sigma](u;B_0)$$

---

8) For half-integer p-period characteristics $(\sigma) = \begin{pmatrix} g_1 & \ldots, & g_p \\ h_1 & \ldots, & h_p \end{pmatrix}_2$ and $(\sigma') = \begin{pmatrix} g_1' & \ldots, & g_p' \\ h_1' & \ldots, & h_p' \end{pmatrix}_2$ the integer $|\sigma',\sigma|$ is defined to be $\operatorname{Exp}\{\pi i \sum_{j=1}^{p} (g_j h_j' + g_j' h_j)\} = \pm 1$. If $|\sigma',\sigma| = 1$ the two period characteristics are said to be syzygetic. See Krazer [15] Chapter VII. We draw attention to this fact about (f_1+f_2) and $(2f_1)$ since we will use it in the characterizations given in Part III of this paper.

$E(u)$ is the constant one by Lemma 1, Part II. $(1/4)\sigma$ is to be defined by one-fourth of the values of $\int_{x_{01}}^{x_{0j}} du_0$ where the path of integration is restricted to Δ_0, for if we shrink $\partial\Delta_0$ and $\partial\Delta_1$ to points in this case we obtain, in the limit

$$\theta[n' + \sum^{q} a'](0;B_q)\{\theta[{}^0_0](u;B_0')\}^2 =$$

$$\ell_\beta'\{\theta[{}^0_0](u;B_0')\}^2.$$

The dependence of the constants ℓ_β on β seems an interesting problem which we believe is open. The last formula shows the dependence is complicated since the ℓ_β's in the limit, here denoted ℓ_β', are the non-vanishing hyperelliptic thetanulls.[9)]

---

9) For another derivation of formula (9) and a deeper insight into the nature of the ℓ_β's see Fay [13] Chapter 5.

4. Other Applications. In this section we consider applications of Corollary 3, Part I, to certain groups of genus two, three, and five.

If W_1 admits a group of automorphisms, G, isomorphic to $Z_2 \times Z_2$ let $G = \{\phi_1, \phi_2, \phi_3, \phi_4\}$ where $\phi_1 = \text{id}$, $\phi_2 \phi_3 = \phi_4$ and $\phi_2^2 = \phi_3^2 = \phi_4^2 = \phi_1$. Let $W_1/\langle\phi_j\rangle = W_j$ and let $W_0 = W_1/G$. We thus have the following covers of surfaces

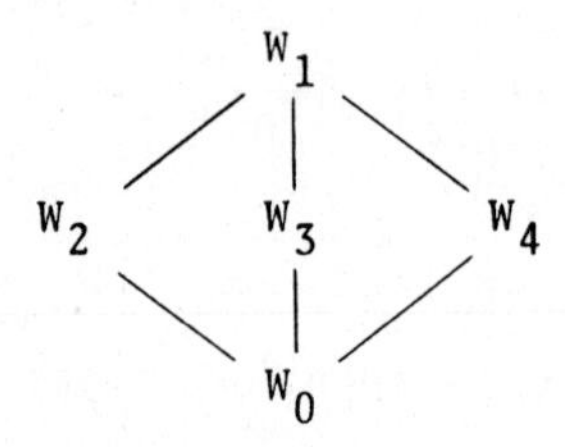

where each line segment corresponds to a two-sheeted cover. If p_i is the genus of W_i, $i = 0,1,2,3,4$ then

$$p_1 + 2p_0 = p_2 + p_3 + p_4 \tag{10}$$ [10].

Let the symbol $(p_1; p_2, p_3, p_4; p_0)$ denote a surface of genus p_1 admitting a $Z_2 \times Z_2$ with the corresponding genera for the quotients. We will consider the three cases

i) $(3;\ 1,\ 1,\ 1;\ 0)$

ii) $(3;\ 0,\ 1,\ 2;\ 0)$

iii) $(5;\ 1,\ 1,\ 3;\ 0)$.

---

10) See Accola [1] p. 478.

A fourth case, iv), will be Z_3 on a surface of genus two. i) (3; 1, 1, 1; 0). In this case $n = 4$. By the Riemann-Hurwitz formula $r = 12$ and $s = 6$. Order the points of X_0 $(X_0 = x_{01} + x_{02} + \dots + x_{06})$ so that:

above $x_{01}, x_{02}, x_{03}, x_{04}$ $W_2 \to W_0$ is branched;

above $x_{01}, x_{02}, x_{05}, x_{06}$ $W_3 \to W_0$ is branched;

above $x_{03}, x_{04}, x_{05}, x_{06}$ $W_4 \to W_0$ is branched.

For the cover $W_1 \to W_0$ $\nu_j = 2$, for $j = 1, 2, \dots, 6$, so

$$V = Z_2 \times Z_2 \times Z_2 \times Z_2 \times Z_2 \times Z_2.$$

The elements of V corresponding to R are

$$\chi^{(1)} \quad (0, 0, 0, 0, 0, 0)$$

$$\chi^{(2)} \quad (1, 1, 1, 1, 0, 0)$$

$$\chi^{(3)} \quad (1, 1, 0, 0, 1, 1)$$

$$\chi^{(4)} \quad (0, 0, 1, 1, 1, 1).$$

Admissible $\beta \in V$ satisfy $\sum_{j=1}^{6} 2\beta_j \equiv 6 \pmod 4$ or $\sum \beta_j$ is odd.*

---

*) $\beta \in V$ will be called <u>admissible</u> if it satisfies the condition of formula (15), Section 5, <u>Part I.</u>

Let $\beta^{(j)}$ be that element of V with a single one in the jth position and zeros elsewhere. Let $\beta^{(i,j,k)}$ have ones only in the i, j and k positions. Modifying $\beta^{(1)}$ by the χ's gives

$$(1, 0, 0, 0, 0, 0)$$

$$(0, 1, 1, 1, 0, 0)$$

$$(0, 1, 0, 0, 1, 1)$$

$$(1, 0, 1, 1, 1, 1).$$

Thus $\sum_{\chi \in R} \max(0, t_\chi) = 1$ for $\beta^{(1)}$ and for any other $\beta^{(j)}$. Now there are 32 admissible β's in all which divide into eight sets of four each when modifications by the χ's are taken into consideration. The $\beta^{(j)}$'s account for 24 of the admissible β's, the other 8 arising from $\beta^{(1,3,5)}$ and $\beta^{(2,4,6)}$. For these last two β's it is seen that $\sum_{\chi \in R} \max(0, t_\chi) = 0$. By Corollary 1, Part I, $\theta[c_1 + c_3 + c_5 - e_1](u;B_1)$ and $\theta[c_2 + c_4 + c_6 - e_1](u;B_1)$ are non-zero at $u = 0$.

By Lemma 6, Part II, e_1 is a 1/4 period. For any j $2c_j - 2e_1 \equiv 2e_1 \not\equiv 0$. Also $(c_1 - e_1) + (c_2 - e_1) \equiv \sum_{j=3}^{6} c_j \equiv 0$ by Lemma 4, Part II. Thus $c_1 - e_1 = -(c_2 - e_1)$. Since $\theta(u;B_1)$ is an even function of u, the vanishing of θ at $c_1 - e_1$ implies the vanishing at $c_2 - e_1$. We now have the following corollary.

<u>Corollary 6</u>: Suppose a surface of genus three, W_1, admits a $(3;\ 1,\ 1,\ 1;\ 0)$. Then there are three quarter periods f_1, f_2, f_3 in $J(W_1)$ so that i) the cyclic groups $\langle f_1\rangle$, $\langle f_2\rangle$, and $\langle f_3\rangle$ are distinct; ii) $2f_1 \equiv 2f_2 \equiv 2f_3$; and iii) $\theta(f_1,B_1) = \theta(f_2,B_1) = \theta(f_3,B_1) = 0$.

ii) $(3;\ 0,\ 1,\ 2;\ 0)$. Again in this case $n = 4$, $r = 12$, and $s = 6$. Analysis as in the previous case will lead only to the one hyperelliptic vanishing property of $\theta(u;B_1)$.
The elliptic-hyperelliptic properties do not show up in analyzing the cover $W_1 \to W_0$ since e_1 is a half-period in this case. Consequently, the following corollary merely records previous results.

<u>Corollary 7</u>: Suppose a surface of genus three admits a $(3;\ 0,\ 1,\ 2;\ 0)$. Then $\theta(u;B_1)$ vanishes to order two at one half-period and to order one at two quarter periods as in Corollary 5.

Since a surface of genus three which admits a $Z_2 \times Z_2 \times Z_2$ necessarily admits a $(3;\ 1,\ 1,\ 1;\ 0)$ and is hyperelliptic[11)] we obtain the following

<u>Corollary 8</u>: Suppose a surface of genus three admits a group of automorphisms isomorphic to $Z_2 \times Z_2 \times Z_2$. The corresponding theta function vanishes to order two at one half-period and vanishes to order one at three quarter-periods as in Corollary 6.

iii) $(5;\ 1,\ 1,\ 3;\ 0)$. By the Riemann-Hurwitz formula $r = 16$ and so $s = 8$. Order the points of X_0 so that

---

11) See Accola [1] p. 478.

above $x_{01}, \dots, x_{04}$ $W_2 \to W_0$ is branched

above $x_{05}, \dots, x_{08}$ $W_3 \to W_0$ is branched.

$W_4 \to W_0$ is branched over each point of X_0. $\nu_j \equiv 2$ for $j = 1,2, \dots, 8$ and $V = Z_2 \times \dots \times Z_2$ (8 times). The elements in V corresponding to R are

$$\chi^{(1)} \quad (0, 0, 0, 0, 0, 0, 0, 0)$$

$$\chi^{(2)} \quad (1, 1, 1, 1, 0, 0, 0, 0)$$

$$\chi^{(3)} \quad (0, 0, 0, 0, 1, 1, 1, 1)$$

$$\chi^{(4)} \quad (1, 1, 1, 1, 1, 1, 1, 1)$$

Admissible β's satisfy $\sum 2\beta_j \equiv 8 \pmod 4$; that is, $\sum \beta_j$ is even. Let $\beta^{(0)}$ be the zero element of V and let $\beta^{(i,j)}$ be the element of V with the only ones in the i and j positions. Since e_1 is a half-period in this case it is seen that there are seven β's which yield vanishing of the theta function at half-periods to order two, namely
$\beta^{(0)}, \beta^{(11)}, \beta^{(12)}, \beta^{(13)}, \beta^{(45)}, \beta^{(46)}, \beta^{(47)}$. Since
$\beta^{(0)} + \beta^{(11)} + \beta^{(12)} + \beta^{(13)} = \beta^{(0)} + \beta^{(45)} + \beta^{(46)} + \beta^{(47)}$ we obtain the following.

<u>Corollary 9</u>: Suppose W_1 is a surface of genus five admitting a $(5; 1, 1, 3; 0)$. Then $\theta(u;B_1)$ vanishes to order two at seven distinct half periods f_i, $i = 0,1,2, \dots, 6$ where

$$0 \equiv f_0 + f_1 + f_2 + f_3 \equiv f_0 + f_4 + f_5 + f_6.$$

iv) Suppose W_1 is a Riemann surface of genus two admitting a cyclic group of order three. By the Riemann-Hurwitz formula we see that the genus of W_1/Z_3 is zero, $r = 8$ and $s = 4$. Each $\nu_j = 3$. The requirement $\sum_{j=1}^{4} (n/\nu_j)\chi_j \equiv 0 \pmod 3$ means that the images of R in V $(= Z_3 \times Z_3 \times Z_3 \times Z_3)$ are as follows

$$\chi^{(1)} \quad (0, 0, 0, 0)$$

$$\chi^{(2)} \quad (1, 1, 2, 2)$$

$$\chi^{(3)} \quad (2, 2, 1, 1)$$

for a suitable ordering of the points of X_0. Admissible β's satisfy $\sum\beta_j \equiv 4 \equiv 1 \pmod 3$. Let $\beta^{(j)}$ be the β with one in the jth position and zeros elsewhere. $\beta^{(1)}$ modified by R gives

$$(1, 0, 0, 0)$$

$$(2, 1, 2, 2)$$

$$(0, 2, 1, 1).$$

Thus for $\beta^{(j)}$ $\sum_{\chi \in R} \max(0, t_\chi) = 1$, $j = 1,2,3,4$. Other permissible β's give $\sum_{\chi \in R} \max(0, t_\chi) = 0$. For example if $\beta = (1, 2, 1, 0)$ then modifying by R gives $(1, 2, 1, 0)$, $(2, 0, 0, 2)$ and

(0, 1, 2, 1). Thus $\theta(u;B_1)$ vanishes at $c_j - e_1$ for $j = 1,2,3,4$. However, $(c_1 - e_1) + (c_2 - e_2) \equiv c_1 + c_2 + \sum c_j \equiv 0$. Also $6(c_j - e_1) \equiv 0$ and $3(c_j - e_1) = -3e_1$ independently of j.

<u>Corollary 10</u>: Suppose W_1 is a surface of genus two admitting a group of automorphisms isomorphic to Z_3. Then there are two one-sixth periods f_1, f_2 so that i) $\langle f_1 \rangle \neq \langle f_2 \rangle$, ii) $3f_1 \equiv 3f_2$ and iii) $\theta(f_1, B_1) = \theta(f_2, B_1) = 0$.

5. <u>Closing Remarks</u>. In the five corollaries of Section 4 the number of vanishing properties is equal to $(3p_1 - 3) - (s - 3) = 3p_1 - s$ which is the codimension in Teichmüller space of the locus of surfaces admitting the particular group.

One can extend the type of result of Corollary 9 to non-hyperelliptic surfaces of genus five admitting groups isomorphic to $Z_2 \times Z_2 \times Z_2$ and $Z_2 \times Z_2 \times Z_2 \times Z_2$. These extensions are not, however, direct corollaries of Theorem 1, Part I. To see how this might happen consider Corollary 7. The elliptic-hyperelliptic vanishing properties come from looking at the cover $W_1 \to W_3$ $(= W_1/\langle\phi_3\rangle)$ rather than by looking at the cover $W_1 \to W_0$.

In general, as the group G becomes more complicated fewer vanishing properties show up by applying Theorem 1 directly to the cover $W_1 \to W_1/G$. For large G there are, in fact, more vanishing properties since G will have more cyclic subgroups. As an example, the reader may wish to check that there are no vanishing properties revealed by applying Theorem 1 directly to a $Z_2 \times Z_2$ on W_1 where p_1 is even and $p_0 = 0$, simply because there are no admissible β's.

Theorem 1 of Part I seems of most use for cyclic groups. An approach to more complicated groups can be seen by considering Corollary 9, (5; 1, 1, 3; 0). The two elliptic-hyperelliptic covers $W_1 \to W_2$ and $W_1 \to W_3$ each give rise to a set of four zeros of order two at half-periods. However, these two sets have one zero in common so there are seven which satisfy the conditions of Corollary 9.[12] Also in Corollary 6 any two of the three quarter-periods correspond to one of the subcovers of order two. However, only in exceptional circumstances like these two corollaries does a direct application of Theorem 1, Part I, give a complete picture. There is needed a general procedure for deciding the intersection properties of sets of zeros obtained by applying Theorem 1 to cyclic subgroups of G.

The same kind of problem can arise in a more general context. Consider a $(4p - 3; 2p - 1, 2p - 1, 2p - 1; p)$; that is, a surface of genus $4p - 3$ admitting an unramified $Z_2 \times Z_2$. Suppose that a canonical homology basis can be chosen on W_0 so that $\ker \underline{a} = \langle \omega_1, \omega_2 \rangle$ where ω_1 and ω_2 are half periods with period characteristics $\begin{pmatrix} 0 & 0 & 0 & \cdots & 0 \\ 1 & 0 & 0 & \cdots & 0 \end{pmatrix}_2$ and $\begin{pmatrix} 0 & 0 & 0 & \cdots & 0 \\ 0 & 1 & 0 & \cdots & 0 \end{pmatrix}_2$. Then Corollary 2 of Theorem 1, Part I yields

$$E(u)\theta[e_1]_2(\underline{a}u;B_1) =$$

$$\ell\theta[0]_2(u;B_0)\theta[\omega_1]_2(u;B_0)\theta[\omega_2]_2(u;B_0)\theta[\omega_1 + \omega_2]_2(u;B_0).$$

There are u_{p-2} $(= (4^{p-2} - 2^{p-2})/2)$ odd period characteristics of the type $\begin{pmatrix} 0 & 0 & \varepsilon_1 & \cdots & \varepsilon_{p-2} \\ 0 & 0 & \varepsilon_1' & \cdots & \varepsilon_{p-2}' \end{pmatrix}_2$ so that if g_0 has one of these

---

12) A proof of this last assertion will be omitted. It depends on an examination of the covers more detailed than that presented in this paper.

period characters then $\theta[e_1](u;B_1)$ vanishes to order four at $u = \underline{a}g_0$.

Consequently, unramified non-cyclic groups of order four yield higher order vanishing properties in addition to zeros arising from the subgroups of order two. Thus for an unramified $Z_2 \times Z_2 \times Z_2$ a further problem is the intersection properties of the seven sets of higher order zeros arising from the seven subgroups of order four.

Finally, one should mention that Wirtinger [28] discovered in the two sheeted unramified case further varieties of zeros on $J(W_1)$ which, so far, seem inaccesible by the methods of Theorem 1, Part I, except in isolated instances.

PART III *

1. Introduction. In the third part of this paper we shall show how some of the vanishing properties of theta functions derived in Part II characterize the existence of automorphism groups. As remarked in Part I, Riemann's solution to the Jacobi inversion problem allows one to infer from the existence of certain vanishing properties of the theta function the existence of certain linear series of degree $p - 1$. The problem then is to show how the existence of such linear series leads to the existence of automorphisms. With one exception we shall show only the existence of automorphisms of period two (involutions). This work can be viewed as a generalization of the classical characterizations of hyperelliptic Riemann surfaces of genus three and four and the modern work of H. H. Martens [20] for arbitrary genus. We will characterize elliptic-hyperelliptic surfaces for all genera except four and further p_0-hyperelliptic cases for higher genus. The main tool will be a theorem essentially due to Castelnuovo [6]. However, the p_0-hyperelliptic characterizations for the lowest genera are reduced to the case of higher genus by extensions of a method due to Farkas [9].

---

* Supported by National Science Foundation Grant GP-21191(A2)

For genus three and five it is possible to use the elliptic-hyperelliptic information to characterize the existence of certain elementary abelian two-groups of orders four, eight and sixteen. Also for hyperelliptic surfaces of genus five a characterization is derived confirming the "p - 2 conjecture". The case where a non-involution is characterized is that of a cyclic group of order three for genus two. Finally, for genus five two local characterizations are obtained for loci of Teichmueller space corresponding to two other automorphism groups. In the final section we discuss some open questions.

We now record some known results and notational conventions which are used in this paper. All divisors will be integral unless otherwise indicated. If D is a divisor so that 2D is canonical, D will be called half-canonical. If W_1 is a Riemann surface and G is a finite group of automorphisms the genus of W_1/G will be called the quotient genus of G. If T is an automorphism of finite order the quotient genus of T will mean the quotient genus of $W_1/\langle T\rangle$. Frequently we will refer to "the theta function" of W_1. In order that this make sense we will assume a canonical homology basis to have been chosen and the corresponding first order theta function with zero characteristic will be designated by "the theta function". Statements concerning "the theta function" will not depend on the canonical homology basis chosen.

If W_1 is a closed Riemann surface of genus $p_1 \geq 1$ and a canonical homology basis is chosen then to each of the $2^{2p_1} - 1$ non-zero period-characteristics corresponds one of the $2^{2p_1} - 1$ smooth two-sheeted coverings of W_1 of genus $2p_1 - 1$.

If (σ) is a period characteristic then (σ) defines a representation of the first homotopy (homology) group onto Z_2 whose kernel defines the two-sheeted smooth covering W_σ corresponding to (σ). If $[\varepsilon_1]$ and $[\varepsilon_2]$ are two theta-characteristics so that $[\varepsilon_1] + [\varepsilon_2] = (\sigma)$ then the quotient of theta functions $\theta[\varepsilon_1](u(p))/\theta[\varepsilon_2](u(p))$ $(p \in W_1$ and $u : W_1 \to J(W_1))$ defines a single valued meromorphic function on W_σ. If both characteristics are odd then the function on W_σ is of order $2p_1 - 2$ and gives rise to special vanishing properties for the theta function of W_σ. This is the fundamental observation of Farkas [9].

If (σ) and (τ) are distinct period-characteristics then there is a four-sheeted cover of W_1, corresponding to the group of period-characteristics $\langle(\sigma),(\tau)\rangle$, on which a fixed-point-free four-group of automorphisms operates whose quotient is W_1 and for which W_σ, W_τ, and $W_{\sigma\tau}$ are the quotients modulo the three cyclic groups of order two. The methods of Farkas extend to this situation and we will make use of this in characterizing p_0-hyperelliptic surfaces.

For hyperelliptic surfaces we need the following facts. If $W_1 \to W_0$ is a covering of closed Riemann surfaces then if W_1 is hyperelliptic so is W_0. For hyperelliptic surfaces the Weierstrass points are the fixed points for the hyperelliptic involution. Also if D is a half-canonical divisor on a hyperelliptic surface then the fixed points for the linear series determined by D are Weierstrass points. Finally, for surfaces of genus four or more a surface cannot be hyperelliptic and elliptic-hyperelliptic at the same time.

For surfaces of genus five we will need the fact that elliptic-hyperelliptic involutions commute and so generate elementary abelian two-groups ([1] p. 480). Also elliptic-hyperelliptic involutions commute with fixed-point-free involutions for genus five. Since this last fact does not seem to be in the literature we include a proof in the appendix to section 5 of this paper.

Finally we note that for surfaces of genus p_1, an involution with quotient genus p_0 is unique if

$$p_1 > 4p_0 + 1. \quad \text{([1] p. 479 or [3] p. 321)}$$

Such an involution is, therefore, central in the full group of automorphisms, and it will be called <u>strongly branched</u>. This generalizes the hyperelliptic situation, $p_0 = 0$.

2. Castelnuovo's method and p_0-hyperellipticity. The following lemma is an adaptation of a theorem of Castelnuovo ([6], or [7] p. 294).[1)]

Lemma 1: Let g^r_{p-1} be a simple half-canonical linear series on a Riemann surface W of genus p. If $r \geq 2$ then $p \geq 3r$.

Proof: Use the linear series to give a birational map of W onto a curve C in $\mathbb{P}^r(\mathbb{C})$ of degree n where $p - 1 - n$ is the number of fixed points of g^r_{p-1}. By Clifford's theorem $n \geq 2r + 1$. The quadrics cut out on C a linear series $g^{r_2}_{2n}$ which is part of the canonical series and so $p - 1 \geq r_2$. Fix a hyperplane H_0 so that the points of the hyperplane section, $z_1, z_2, \ldots, z_n$, are in general position in H_0; that is, thru any $r - 1$ of the points there is a hyperplane containing no further z's. Thus thru any $2(r - 1)$ points one can find two hyperplanes (that is, a quadric) containing no further z's.

Let $z_1, \ldots, z_\nu$ be a set of z's so that i) any quadric thru $z_1, \ldots, z_\nu$ passes thru the remaining $n - \nu$ z's and ii) the points $z_1, \ldots, z_\nu$ impose independent conditions on the quadrics. By the preceding argument $2r - 1 \leq \nu \leq n$.

Now consider the non-fixed points of the linear series $|g^{r_2}_{2n} - z_1 - z_2 - \ldots - z_\nu|$. It is a $g^{r_1}_{n}$ and contains g^r_n. Thus $r \leq r_1 \leq r_2 - \nu$ or $p - 1 \geq r_2 \geq r + \nu \geq 3r - 1$. q.e.d.

---

1) The author thanks Alan Landman for valuable discussions concerning the material in this section.

In Corollaries 2 and 3 of Part II of this paper, the highest order vanishing properties of p_0-hyperelliptic theta-functions are recorded. If p_1 is odd and W_1 is p_0-hyperelliptic then the theta function vanishes at 4^{p_0} half-periods to order $(p_1 + 1)/2 - p_0$. If p_1 is even then the theta function vanishes at $(2p_1 + 2 - 4p_0)4^{p_0}$ half-periods to order $p_1/2 - p_0$. We now use Lemma 1 to show that if p_1 is large and p_0 small one of these vanishing properties suffices to insure q_0-hyperellipticity where $q_0 \leq p_0$.

<u>Theorem 1</u>: Suppose W_1 is a closed Riemann surface of odd genus p_1. Suppose W_1 admits a complete half-canonical $g^{(p_1-1)/2-p_0}_{p_1-1}$ where $p_1 \geq 6p_0 + 5$. Then W_1 is q_0-hyperelliptic where $q_0 \leq p_0$.

<u>Proof</u>: If the linear series is simple then by Lemma 1

$$p_1 \geq 3[(p_1 - 1)/2 - p_0]$$

or

$$p_1 \leq 6p_0 + 3.$$

Thus $g^{(p_1-1)/2-p_0}_{p_1-1}$ is composite and there is a covering $b : W_1 \to W_0$ of t sheets and a complete simple linear series $g^{(p_1-1)/2-p_0}_{n/t}$ on W_0 which lifts via b to the original series on W_1. n is the number of non-fixed points of the original linear series. If $t \geq 3$ then

$$(p_1 - 1)/2 - p_0 \leq n/t \leq (p_1 - 1)/t \leq (p_1 - 1)/3$$

or

$$p_1 \leq 6p_0 + 1$$

a contradiction. Thus $t = 2$. If $g^{(p_1-1)/2-p_0}_{n/2}$ on W_0 is special then by Clifford's theorem

$$(p_1 - 1)/2 - [(p_1 - 1) - 2p_0] \geq n/2 - (p_1 - 1) + 2p_0 \geq 0$$

or

$$p_1 \leq 4p_0 + 1$$

again a contradiction. Thus $g^{(p_1-1)/2-p_0}_{n/2}$ is not special and so by the Riemann-Roch theorem

$$q_0 = n/2 - [(p_1 - 1)/2 - p_0] \leq p_0 \qquad \text{q.e.d.}$$

The following immediate corollary will be useful.

Corollary 1: If it is known in Theorem 1 that $q_0 = p_0$ then $g^{(p_1-1)/2-p_0}_{p_1-1}$ is without fixed points and every divisor in the linear series is invariant under the involution of W_1 whose quotient is W_0.

As in the hyperelliptic case, the results for even genus are not as neat. This is because the linear series derived from the highest order vanishing property always has at least one fixed point.

Theorem 2: Suppose W_1 is a Riemann surface of even genus p_1. Suppose W_1 admits a complete half-canonical $g^{(p_1-2)/2-p_0}_{p_1-1}$ where $p_1 \geq 6p_0 + 8$. Then W_1 is q_0-hyperelliptic where $q_0 \leq p_0$.

Proof: If the linear series is simple then by Lemma 1

$$p_1 \geq 3[(p_1 - 2)/2 - p_0]$$

or

$$p_1 \leq 6p_0 + 6$$

a contradiction. As in Theorem 1, there is a map of two sheets $b : W_1 \to W_0$ and a $g_{n/2}^{(p_1-2)/2-p_0}$ on W_0 which is complete and non-special. Thus

$$(p_1 - 1)/2 - (p_1 - 2)/2 + p_0 \geq n/2 - [(p_1 - 2)/2 - p_0] = q_0$$

or

$$q_0 \leq p_0 + 1/2 \qquad \text{q.e.d.}$$

Notice that in Theorems 1 and 2 $p_1 > 4q_0 + 1$ so the two-sheeted covering $b : W_1 \to W_0$ is strongly branched; that is, the involution of W_1 whose quotient is W_0 is central in the full group of automorphisms of W_1.

3. Extensions. To obtain characterizations of p_0-hyperellipticity for genera lower than in the last section we employ Theorem 1, Part I and its Corollary 2.

Let G be a finite subgroup of $J(W_1)$ whose order is n, $G = \{f_1, f_2, \ldots, f_n\}$. Suppose for some half-period η, θ vanishes at each point of finite order $f_j + \eta$ to order $s_j \geq 0$. Let W_2 be the smooth abelian n-sheeted cover of W_1 corresponding to the group G; that is, the map $\underline{a} : J(W_1) \to J(W_2)$ has kernel G. Then W_2 has genus $n(p_1 - 1) + 1$. By Corollary 2, Part I

$$E(u)\theta[e_1](\underline{a}u;B_2) = \ell \prod_{f_j \in G} \theta[f_j](u;B_1)$$

where $2e_1 \equiv 0$ and $\theta[e_1](\cdot, B_2)$ has a zero of order $\sum_{j=1}^{n} s_j$ at the half-period $\underline{a}\eta$. Thus W_2 admits a complete, half-canonical series of dimension $(\sum s_j) - 1$. To state this a little differently, note that the vanishing of the theta function on $J(W_1)$ to order s_j at $f_j + \eta$ corresponds to a linear series on W_1 of dimension $r_j = s_j - 1$ and degree $p_1 - 1$ if $s_j \geq 1$. If all $s_j \geq 1$ then the dimension of the half-canonical series on W_2 is $(n - 1) + \sum r_j$. The addition of $(n - 1)$ to what might be expected makes the following extensions of Theorems 1 and 2 possible.[2)]

Corollary 2: Suppose W_1 admits two distinct half-canonical series $g^{r_1}_{p_1 - 1}$ where

---

2) This technique was used in [4] p. 17.

a) $r_1 = (p_1 - 1)/2 - p_0$ if p_1 is odd and $p_1 \geq 6p_0 + 1$

b) $r_1 = (p_1 - 2)/2 - p_0$ if p_1 is even and $p_1 \geq 6p_0 + 4$.

Then W_1 is q_0-hyperelliptic for some $q_0 \leq p_0$.

Proof: The previous discussion can be applied with $n = 2$, for if the characteristics corresponding to the two half-canonical series are η_1 and η_2 let $G = \{0, \eta_1 + \eta_2\}$. W_2 then admits a half-canonical $g^{r_2}_{2p_1-2}$ $(2p_1 - 2 = p_2 - 1)$ where:

in case a) $r_2 = p_1 - 2p_0$ and

in case b) $r_2 = p_1 - 1 - 2p_0$.

Thus in case a) $r_2 = (p_2 - 1)/2 - (2p_0 - 1)$ and $p_2 > 6(2p_0 - 1) + 5$. Consequently, by Theorem 1 W_2 is q_1-hyperelliptic where $q_1 \leq 2p_0 - 1$. In case b) $r_2 = (p_2 - 1)/2 - 2p_0$ and again $p_2 > 6(2p_0) + 5$. Again by Theorem 1, W_2 is q_1-hyperelliptic where $q_1 \leq 2p_0$. Thus W_2 admits two involutions, one whose quotient is W_1 and the second whose quotient is a surface of genus q_1. Since the latter involution commutes with the former, they generate a four-group on W_2 whose quotient W_0 has genus q_0 satisfying

$$2q_0 - 1 \leq q_1 \quad \text{or} \quad q_0 \leq p_0.$$

Since W_1 is a two-sheeted cover of W_0, the theorem is proven.

q.e.d.

Corollary 3: Suppose W_1 admits four distinct half-canonical series $g^{r_1}_{p_1-1}$ so that the corresponding four half-periods sum to zero in $J(W_1)$ where

a) $r_1 = (p_1 - 1)/2 - p_0$ if p_1 is odd and $p_1 \geq 6p_0 - 1$

b) $r_1 = (p_1 - 2)/2 - p_0$ if p_0 is even and $p_1 \geq 6p_0 + 2$.

Then W_1 is q_0-hyperelliptic for some $q_0 \leq p_0$.

Proof: Let the half-periods in the statement of the corollary be η_1, η_2, η_3, and η_4 where $\eta_1 + \eta_2 + \eta_3 = \eta_4$. If we let the group G, of the discussion preceding Corollary 2, be the four-group $\{0, \eta_1 + \eta_2, \eta_1 + \eta_3, \eta_2 + \eta_3\}$ and let η_1 be the half-period by which G is translated, we see that the condition $\sum_{j=1}^{4} \eta_j \equiv 0$ allows us to apply the conclusions of that discussion. We obtain a smooth four-sheeted abelian cover W_2 of W_1 on which there is a half-canonical series $g^{r_2}_{p_2-1}$ where $(p_2 - 1 = 4(p_1 - 1))$:

in case a) $r_2 = 2p_1 - 2 - 4p_0 + 3$ and

in case b) $r_2 = 2p_1 - 4 - 4p_0 + 3$.

Thus in case a) $r_2 = (p_2 - 1)/2 - (4p_0 - 3)$ and $p_2 = 4p_1 - 3 > 6(4p_0 - 3) + 5$. Consequently, by Theorem 1 W_2 is q_1-hyperelliptic where $q_1 \leq 4p_0 - 3$. In case b) $r_2 = (p_2 - 1)/2 - (4p_0 - 1)$ and $p_2 > 6(4p_0 - 1) + 5$. Consequently, W_2 is q_1-hyperelliptic where $q_1 \leq 4p_0 - 1$

Thus W_2 admits a four-group of automorphisms whose quotient is W_1 and an involution whose quotient is a surface of genus q_1. Since the latter involution commutes with the involutions of the four-group, all together they generate an elementary abelian group of order eight whose quotient W_0 has genus q_0 satisfying

$$4q_0 - 3 \leq q_1 \quad \text{or} \quad q_0 \leq p_0.$$

Since W_1 is a two-sheeted cover of W_0 the theorem is proven.

q.e.d.

The last corollary squeezes about as much out of this method as seems possible. We omit the proof since it is entirely analogous to that of the previous corollary.

Corollary 4: Suppose W_1 admits eight distinct half-canonical series $g^{r_1}_{p_1-1}$ so that the eight corresponding half-periods are a translate (by a half-period) of a subgroup of $J(W_1)$ of order eight where

a) $r_1 = (p_1 - 1)/2 - p_0$ if p_1 is odd and $p_1 \geq 6p_0 - 3$

b) $r_1 = (p_1 - 2)/2 - p_0$ if p_1 is even and $p_1 \geq 6p$.

Then W_1 is q_0-hyperelliptic for some $q_0 \leq p_0$.

For $p_0 = 0$ Theorems 1 and 2 and Corollary 2 give the known characterizations for hyperelliptic surfaces. For $p_0 \geq 2$ it is easy to see from the table, Corollary 1, Part II, that the theta function vanishes appropriately on a coset of a $Z_2 \times Z_2 \times Z_2$ in $J(W_1)$.

For $p_0 = 1$, p_1 even, $p_1 \geq 8$ or p_1 odd, $p_1 \geq 5$ again an appropriate four-group can be found to allow Corollary 3 to apply. Also $p_0 = 1$ and $p_1 = 6$ an appropriate $Z_2 \times Z_2 \times Z_2$ can be found to allow Corollary 4 to apply. However, for $p_0 = 1$, $p_1 = 3$ Corollary 4 does not apply. We treat this case in Section 6 where we derive additional information.

We can summarize the results of the last two sections by saying that p_0-hyperellipticity is characterized by the vanishing properties of the theta function at half-periods for Riemann surfaces of genus p_1 where for odd p_1 we have $p_1 \geq 6p_0 - 3$, $p_1 \geq 5$ and for even $p_1 \geq 6p_0$, $p_1 \geq 6$.

It's worth remarking that the two cases of non-strongly branched involutions characterized by these methods are $p_1 = 5$, $p_0 = 1$ and $p_1 = 9$, $p_0 = 2$. The elliptic-hyperelliptic case for genus five is treated in section 5 of this paper.

4. The $p - 2$ conjecture for $p = 5$. The methods of sections 2 and 3 allow us to give the following characterization of hyperelliptic surfaces of genus five. The title of this section is motivated by the fact that for genus p the hyperelliptic locus in Teichmueller space has codimension $p - 2$.

Corollary 5: Let W_5 be a Riemann surface of genus 5. Suppose the theta function vanishes to order two at three half-periods η_1, η_2 and η_3 and vanishes to order one at $\eta_1 + \eta_2 + \eta_3$. Then W_5 is hyperelliptic.

Proof: If one of the corresponding g^1_4's has two fixed points the hyperelliptic g^1_2 is evident. We now show that no g^1_4 can have precisely one fixed point by showing that a W_5 admitting a g^1_3 without fixed points can admit at most one half-canonical g^1_4. Necessarily the g^1_3 is unique, the surface is not hyperelliptic, and any canonical divisor containing two points of a divisor in g^1_3 must contain the third.

Assuming W_5 has a g^1_3 let g^1_4 be a half-canonical divisor and let $x + y + z$ be three distinct points constituting a divisor of g^1_3, no point being a fixed point of g^1_4. Let D_x and D_y be two divisors in g^1_4 containing x and y respectively. Being canonical, $D_x + D_y$ must contain z, say z is in D_x. If D is a divisor of g^1_4 containing no point of $x + y + z$ we see that D_x also contains y since $D + D_x$ is canonical. Thus $D_x = x + y + z + t$ where t is the fourth point; that is $g^1_4 = g^1_3 + t$. If $g^1_3 + s$ is a second half-canonical g^1_4 then $2s \equiv 2t$ and so $s = t$.

Thus, if our surface is not hyperelliptic, none of the half-canonical g^1_4's has a fixed point. By the discussion at the beginning of section 3 there is a four-sheeted abelian cover, W_{17}, of W_5 of genus 17 admitting a complete half-canonical g^6_{16} without fixed points. By Lemma 1 the g^6_{16} is composite and so W_{17} is a two-sheeted cover of a surface admitting a complete g^6_8 without fixed points. This surface must have genus two and will be denoted W_2. Thus W_{17} admits a four-group whose quotient is W_5 and a central involution whose quotient is W_2. These generate an elementary abelian group of order eight, G, whose quotient we will call W_0. We wish to show that the genus of W_0 is zero since W_5 is a two-sheeted cover of W_0. But W_2 admits a four-group whose quotient is W_0. Applying formula (10) of Part II shows that any four-group on a surface of genus two must be a (2;1,1,0;0). Thus W_0 has genus zero, and W_5 is hyperelliptic after all. q.e.d.

5. <u>Elliptic-hyperelliptic surfaces of genus five</u>. Since a surface of genus five can admit several elliptic-hyperelliptic involutions, it is important to establish a one-to-one correspondence between elliptic-hyperelliptic involutions and sets of four half-periods where according to Corollary 3 the theta function vanishes to order two. This will allow us to characterize elementary abelian two-groups generated by several elliptic-hyperelliptic involutions.

<u>Lemma 2</u>: Let $b : W_5 \to W_1$ be a two-sheeted cover of surfaces of genera five and one respectively. Then there are precisely four half-canonical g^1_4's on W_5 which arise from lifting g^1_2's from W_1. These g^1_4's give rise to the vanishing properties of Corollary 3, Part II, via the Riemann vanishing theorem.

<u>Proof</u>: Let $|D_5| = g^1_4$ on W_5, where g^1_4 is the lift of a g^1_2 from W_1 and g^1_4 is half-canonical. If X_5 is the divisor of branch points of the cover $W_5 \to W_1$ then X_5 is canonical and $2D_5 \equiv X_5$. If f_5 is a function on W_5 whose divisor is $2D_5 - X_5$ then f_5 is either symmetric or anti-symmetric with respect to the involution T of W_5 whose quotient is W_1 because the divisor of f_5 is invariant under T. Since the poles of f_5 are simple at the branch points f_5 must be anti-symmetric. Thus $(f_5)^2$ is symmetric and is the lift of a function f_1 on W_1 where f_1 has simple poles at $X_1 (= b(X_5))$ and fourth order zero at D_1 where $|D_1| = g^1_2$ and D_5 is the lift of D_1 via b. Thus

$$4D_1 \equiv X_1. \tag{1}$$

The number of g^1_2's satisfying formula (1) is sixteen, the number of quarter-periods on the Jacobian of W_1 (which is W_1 itself since W_1 is a torus).

Now suppose D_1 and D_1' lift to W_5 to be D_5 and D_5', half-canonical divisors. Then $2D_5 \equiv 2D_5'$ and it follows that $2D_1 \equiv 2D_1'$ on W_1. Thus the D_1's which satisfy formula (1) <u>and</u> lift to half-canonical divisors on W_5 correspond to the half-periods on (the Jacobian of) W_1, and so are precisely four in number.

That these g^1_4's are the ones derived from Corollaries 4 and 1 of Part II follows from the fact that the map $\underline{a} : J(W_1) \to J(W_5)$ was defined by lifting divisors from W_1 to W_5.

<u>Theorem 3</u>: Let W_5 be a non-hyperelliptic Riemann surface of genus five. Suppose there are four even half-integer theta-characteristics $[\eta_k]$, $k = 1,2,3,4$ so that $\eta_1 + \eta_2 + \eta_3 = \eta_4$ and $\theta[\eta_k](u)$ vanishes to order two at $u = 0$. Then there is an elliptic-hyperelliptic involution T, $W_1 = W_5/\langle T\rangle$, and the four vanishing properties arise from the cover $W_5 \to W_1$ as in Corollary 4, Part II.

<u>Proof</u>: In the proof of Corollary 3 we showed that W_5 admits a smooth four-sheeted abelian cover W_{17}, of genus 17, which admits an involution T' which is hyperelliptic or elliptic-hyperelliptic. If G' is the four group on W_{17} whose quotient is W_5 then T' and G' generate on W_{17} an elementary abelian two-group of order eight whose quotient W_1 is covered in two sheets by W_5. Since W_5 is not hyperelliptic the genus of W_1 must be one and T' must be an elliptic-hyperelliptic involution of W_{17}. Since the four g^1_4's on W_5 lift to the same complete g^7_{16} on W_{17} and since T' is elliptic-hyperelliptic the conditions of Corollary 1 to Theorem 2, this part, hold and so each divisor of g^7_{16} is invar-

iant under T'. If D_4 is a divisor in a g^1_4 on W_5 then it lifts to a divisor D_{16} on W_{17} which is invariant under G' as well. Thus this D_{16} is invariant under all of G and so D_4 must be invariant under T, the involution of W_5 whose quotient is W_1. Thus the four g^1_4's arise from lifting g^1_2's from W_1 to W_5 and the proof is completed by an appeal to Lemma 2.

We now investigate how the different sets of four half-integer theta-characteristics intersect.

<u>Definition</u>: If T is an elliptic-hyperelliptic involution on a surface of genus five then the set of four half-integer theta-characteristics associated with T will be called a <u>T-family</u>.

<u>Theorem 4</u>: If T_1 and T_2 are two distinct elliptic-hyperelliptic involutions for a surface of genus five, the corresponding T-families have precisely one theta-characteristic in common. Three distinct T-families have no theta-characteristics in common.

<u>Proof</u>: Suppose T_1 and T_2 are elliptic-hyperelliptic involutions for W_5. Then $\langle T_1, T_2 \rangle$ (=G) is a four-group (5;1,1,3;0). Let ${}_1W_1 = W_5/\langle T_1 \rangle$, ${}_2W_1 = W_5/\langle T_2 \rangle$, $W_3 = W_5/\langle T_1 T_2 \rangle$, and $W_0 = W_5/G$. Then we have the following diagram of covers corresponding to G:

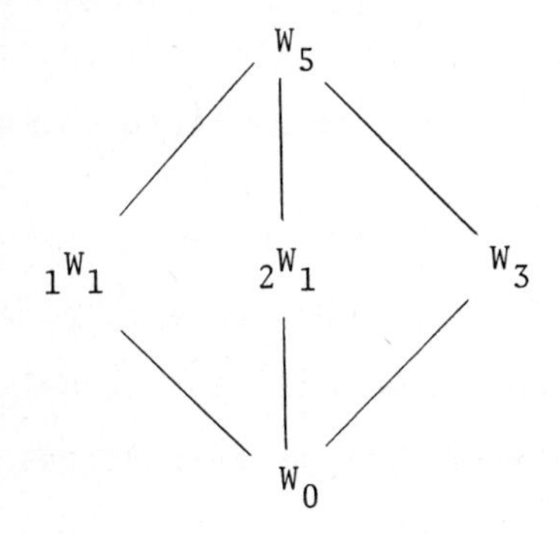

W_3 is hyperelliptic and the hyperelliptic g^1_2 on W_3 arises from lifting the unique g^1_1 on W_0. Thus the g^1_4 on W_5 arising from g^1_2 on W_3 is half-canonical and invariant under T_1 and T_2. Consequently, this g^1_4 corresponds to a theta-characteristic in each T-family. Conversely, a theta-characteristic in both T-families corresponds to a g^1_4 whose divisors are invariant under T_1 and T_2, and therefore, under all of G. Thus it comes from the unique g^1_1 on W_0. The first statement of Theorem 4 is proven.

Now suppose W_5 admits three elliptic-hyperelliptic involutions T_1, T_2, and T_3. A g^1_4 corresponding to a theta-characteristic in each T-family must have divisors of degree eight since each divisor would be invariant under the group of order eight generated by the T's. This contradiction completes the proof. q.e.d.

Using Theorem 4 we can now characterize those surfaces of genus five admitting elementary abelian two-groups generated by two, three, and four-elliptic-hyperelliptic involutions. Notice that Theorem 4 implies that a surface of genus five can admit at most five elliptic-hyperelliptic involutions.

Corollary 6: A non-hyperelliptic Riemann surface of genus five admits a group of automorphisms isomorphic to $Z_2 \times Z_2$ generated by two elliptic-hyperelliptic involutions if and only if $\theta[\eta](u)$ vanishes to order two at $u = 0$ for seven theta-characteristics, $[\eta]$, corresponding to two T-families.

Corollary 7: A non-hyperelliptic Riemann surface of genus five admits a group of automorphisms isomrophic to $Z_2 \times Z_2 \times Z_2$ generated by three elliptic-hyperelliptic involutions if and only if $\theta[\eta](u)$ vanishes to order two at $u = 0$ for nine theta-characteristics, $[\eta]$, corresponding to three T-families.

<u>Corollary 8</u>: A (non-hyperelliptic) Riemann surface of genus five admits a group of automorphisms isomorphic to $Z_2 \times Z_2 \times Z_2 \times Z_2$ if and only if $\theta[\eta](u)$ vanishes to order two at $u = 0$ for ten theta-characteristics, $[\eta]$ corresponding to four T-families.

<u>Proof</u>: By Theorem 4 and the inclusion-exclusion counting principle three T-families contain

$$3 \cdot 4 - 3 \cdot 1 + 1 \cdot 0 = 9$$

theta-characteristics, and four T-families contain

$$4 \cdot 4 - 6 \cdot 1 + 4 \cdot 0 - 1 \cdot 0 = 10$$

theta-characteristics. q.e.d.

In the last three corollaries the codimensions in Teichmueller space for genus five of the loci of surfaces admitting the particular groups is the number of conditions given. Consequently, the various loci are seen to be complete intersections.

The surfaces of Corollary 8 deserve further comment. A surface of genus five admitting a $Z_2 \times Z_2 \times Z_2 \times Z_2$ $(=G)$ necessarily has G generated by four elliptic-hyperelliptic involutions since hyper-elliptic surfaces admit elementary abelian two groups of order at most eight. If T_k, $k = 1,2,3,4$ generate G then T_jT_k and $T_jT_kT_\ell$ all have quotient genus three while $T_5(= T_1T_2T_3T_4)$ is again elliptic-hyperelliptic. Thus there is a fifth T-family among the ten theta-characteristics. This is, however, consistent with Theorem 4 and Corollary 8 since, by the inclusion-exclusion principle

$$5 \cdot 4 - 10 \cdot 1 + 10 \cdot 0 - 5 \cdot 0 + 1 \cdot 0 = 10.$$

These surfaces are known as Humbert's surfaces and have been studied in [8], [14], and [27]. They have the property that a basis of abelian differentials of the first kind can be achieved by five reducible abelian differentials which are, in fact, elliptic. These five differentials are the elliptic integrals lifted from the five quotient surfaces for subgroups of order eight generated by T_j, $T_k T_\ell$, $T_k T_m$ for the five choices of T_j. Such a subgroup has only one elliptic-hyperelliptic involution, T_j, and thus six involutions of quotient genus three. It follows from formula (6) [1] that the quotient genus of such a group of order eight must be one. That the five differentials are linearly independent follows from the fact that they correspond to five different characters of G. (See [2], pp. 600-601.)

We also remark, without proof, that there are no further similar vanishing properties of the theta function for Humbert's surfaces, for an eleventh half-canonical g^1_4 would imply a sixth elliptic-hyperelliptic involution or a hyperelliptic involution.

Appendix: We will need the following proposition in the next section so we include it here.

Proposition: Let W be a closed Riemann surface of genus five. Let T be an elliptic-hyperelliptic involution and let S be a fixed-point-free involution. Then T and S commute.

<u>Proof</u>: Let D_n be the dihedral group generated by T and S of order 2n. We wish to show that n is two. If n were odd all reflections in D_n would be conjugate and thus T and S would have the same quotient genus; consequently, n is even. Let R(= TS) generate the cyclic group of order n. Then T and $TR^2(= R^{-1}TR)$ are conjugate and so T and TR^2 are elliptic-hyperelliptic involutions. Therefore T and TR^2 commute and so R^2 is central in D_n. Since the only central element of $D_n (n > 2)$ is $R^{n/2}$ we see that n = 2 or n = 4.

If n = 4 then $\langle T,TR^2\rangle$ is a (5;1,1,3;0) and so R^2 is a fixed-point-free involution of quotient genus three. Consequently $\langle R\rangle$ is fixed point free of quotient genus two and $\langle S,SR^2\rangle$ is a (5;3,3,3;2). Now consider $D_4/\langle R^2\rangle$ acting on $W_5/\langle R^2\rangle$. The three groups of order four on W_5, $\langle T,TR^2\rangle$, $\langle R\rangle$, and $\langle S,SR^2\rangle$ reduce to commuting involutions on $W_5/\langle R^2\rangle$ of quotient genera zero, two, and two. But this is impossible by formula (10) Part II and so $n \neq 4$. The proof is complete.

6. Elliptic-hyperelliptic surfaces of genus three. The methods of this paper become more circuitous as the genus decreases, and this section is no exception. However, any surface of genus three admitting an involution is hyperelliptic, elliptic-hyperelliptic, or both. Consequently, we will be able to completely characterize elementary abelian two-groups on surfaces of genus three by vanishing properties of the theta function. How these characterizations might lead to characterizations of non-abelian automorphism groups for genus three is indicated in the closing remarks, Section 9.

The tasks of this section are to show that the conclusions of Corollaries 5,6,7 and 8 of Part II characterize the surfaces admitting the indicated automorphism groups. Since we will assume the classical characterization of hyperellipticity in terms of the vanishing of the theta function at one half-period to order two, we need only consider converses to Corollaries 5 and 6.

Theorem 5: Let W_3 be a Riemann surface of genus three. Suppose that there are two quarter-periods f_1 and f_2 so that i) $f_1 \not\equiv \pm f_2$; ii) $2f_1 \equiv 2f_2 \not\equiv 0$; iii) $|(2f_1),(f_1 + f_2)| = 1$; and iv) the theta function vanishes at f_1 and f_2. Then W_3 is elliptic-hyperelliptic.

Proof: Let $\sigma \equiv 2f_1 \equiv 2f_2 \not\equiv 0$; σ is a half-period. Let W_5 be the smooth two-sheeted cover of W_3 corresponding to σ and S be the involution of W_5 with W_3 as quotient. We wish to show that W_5 admits an elliptic-hyperelliptic involution T that commutes with S so that the quotient genus of $\langle S,T\rangle$ is one.
This will prove the theorem and, as a bonus, give information about (5;3,3,1;1)'s leading to a local characterization in Section 8.

By a classical technique ([15] p. 280) we may assume, without loss of generality, that the canonical homology basis on W_3 has been chosen so that $(\sigma) = \binom{0\ 0\ 0}{1\ 0\ 0}_2$, $(f_1 + f_2) = \binom{0\ 0\ 0}{0\ 1\ 0}_2$ and so $(f_1 - f_2) = \binom{0\ 0\ 0}{1\ 1\ 0}_2$. Thus the syzygetic group $G = \langle(\sigma),(f_1 + f_2)\rangle$ consists of these three period characteristics and $\binom{0\ 0\ 0}{0\ 0\ 0}_2$. Now it is easy to see that there is a unique set of four odd theta-characteristics obtained by adding to each element of G the same theta-characteristics; namely $\begin{bmatrix}0&0&1\\0&0&1\end{bmatrix}_2$, $\begin{bmatrix}0&0&1\\1&0&1\end{bmatrix}_2$, $\begin{bmatrix}0&0&1\\0&1&1\end{bmatrix}_2$, and $\begin{bmatrix}0&0&1\\1&1&1\end{bmatrix}_2$ which are obtained by adding $\begin{bmatrix}0&0&1\\0&0&1\end{bmatrix}_2$ to each element of G. We now write these four theta-characteristics as

$$[\eta_1],\ [\eta_1 + \sigma],\ [\eta_2],\ [\eta_2 + \sigma]$$

where $$[\eta_1] = \begin{bmatrix}0&0&1\\0&0&1\end{bmatrix}_2 \quad\text{and}\quad [\eta_2] = \begin{bmatrix}0&0&1\\0&1&1\end{bmatrix}_2 .$$

By a classical theorem ([15] p. 258) there are four other pairs of odd theta-characteristics $[\eta_k]$, $[\eta_k + \sigma]$, $k = 3,4,5,6$. As shown by Farkas the two g^0_2's on W_3 corresponding to a particular pair lift to equivalent g^0_4's on W_5 and give rise to a complete g^1_4 which is half-canonical. Thus the six pairs $[\eta_k]$, $[\eta_k + \sigma]$ give rise to six half-canonical series ${}_kg^1_4$ on W_5 and so give rise to the special vanishing properties of the theta function for a W_5 which admits a fixed point free involution. We now show that the quarter- periods f_1 and f_2 also lead to half-canonical g^1_4's on W_5.

Let $\underline{a} : J(W_3) \to J(W_5)$ be the map defined in Section 2, Part I by lifting divisors. Thus $\underline{a}\sigma \equiv 0$. Since $2f_1 \equiv 2f_2 \equiv \sigma$ on W_3,

$\underline{a}f_j$ is a one-half period in $J(W_5)$. If g^0_2 and $g^0_2{}'$ correspond on W_3 to the vanishing of the theta function at f_1 and $-f_1$, then the lifts of these divisors are equivalent on W_5 since $\underline{a}(f_1) \equiv \underline{a}(-f_1)$ and so give rise to a half- canonical g^1_4. We thus have two more half-canonical g^1_4's arising from the quarter periods which we will denote ${}_7g^1_4$ and ${}_8g^1_4$. Now

$$f_1 + f_2 + \eta_1 + \eta_2 \equiv \binom{0\ 0\ 0}{0\ 1\ 0}_2 + \binom{0\ 0\ 1}{0\ 0\ 1}_2 + \binom{0\ 0\ 1}{0\ 1\ 1}_2 \equiv 0$$

as period characteristics and so

$$\underline{a}f_1 + \underline{a}f_2 + \underline{a}\eta_1 + \underline{a}\eta_2 \equiv 0 \quad \text{in} \quad J(W_5).$$

Consequently, the four ${}_kg^1_4$'s $k = 1,2,7,8$ give rise to theta-characteristics satisfying the hypotheses of Theorem 2, Part III. However, we do not know that W_5 is not hyperelliptic since we do not exclude the possibility that W_3 might be. Nevertheless, by Theorem 2, Part III we can let T_5 be a hyperelliptic or elliptic-hyperelliptic involution on W_5 whose existence is infered by these four vanishing properties.

T_5 commutes with S for if T_5 is hyperelliptic it commutes with any automorphism of W_5 and if it is elliptic-hyperelliptic the result is that of the proposition at the end of section 5. Consequently, $\langle S,T_5\rangle$ is a four group whose quotient genus is zero or one. The proof of the theorem will be complete if we show this quotient genus is one. Therefore, suppose $W_0 = W_5/\langle S,T_5\rangle$ has genus zero. Then the involution T_3 on W_3 induced by T_5 is the hyperelliptic involution of W_3. Let D_3 be the divisor on

W_3 of degree two corresponding to the vanishing of the theta function at the quarter period f_1. Let D_5 be the lift of D_3 to W_5. Then D_5 corresponds to the vanishing of the theta-function on W_5 at a half-period to order two. Therefore D_5 is invariant under T_5 and so D_3 is invariant under T_3. Consequently, D_3 must be half-canonical, that is, f_1 must be a half-period. This contradiction shows that T_3 and therefore T_5 are elliptic-hyperelliptic involutions. q.e.d.

By formula (10), Part II, the four group $\langle S,T_5\rangle$ on W_5 must be a (5;3,3,1;1). We thus obtain the following corollary.

Corollary 9: Let W_5 be a Riemann surface of genus five admitting a (5;3,3,1;1). Then $\theta[\eta_k](u)$ vanishes to order two at $u = 0$ for eight half-periods $[\eta_k]$ where $\eta_1 + \eta_2 + \ldots + \eta_6 \equiv 0$ and $\eta_1 + \eta_2 + \eta_7 + \eta_8 \equiv 0$. The two sets of $[\eta_k]$'s $k = 1,2, \ldots ,6$ and $k = 3,4, \ldots ,8$ correspond to the two covers $W_5 \to W_3$ [3] and the four $[\eta_k]$'s $k = 1,2,7,8$ form a T-family for the two-sheeted cover $W_5 \to W_1$.

We now turn to the converse of Corollary 6, Part II. We could apply Theorem 5 three times to obtain a converse, but to apply Theorem 5 we must know that hypothesis iii) holds in all cases. While this is, in fact, the case we prefer for the sake of brevity the different proof which follows.

---

3) That the second set of six theta-characteristics do, in fact, correspond to the second fixed-point-free involution is not proved, although it is extremely plausible. The only proof we know uses techniques different from those of this paper and is quite tedious. Therefore, we omit the proof.

<u>Theorem 6</u>: Let W_3 be a Riemann surface of genus three. Suppose there are three quarter-periods f_1, f_2 and f_3 so that i) the cyclic groups $\langle f_1\rangle$, $\langle f_2\rangle$ and $\langle f_3\rangle$ are distinct; ii) $2f_1 \equiv 2f_2 \equiv 2f_3$, and iii) $\theta(f_j) = 0$ for $j = 1,2,3$. Then W_3 admits a (3;1,1,1;0).

<u>Proof</u>: Applying the Riemann vanishing theorem, let A_k and B_k denote divisors of degree two satisfying

$$u(A_k) + K \equiv f_k$$

$$u(B_k) + K \equiv -f_k$$

for $k = 1,2,3$. Since $2f_j \equiv \pm 2f_k$ for all j,k we see that the six divisors of degree four $2A_k$, $2B_k$, $k = 1,2,3$ are all part of the same linear series g^r_4 where $r = 1$ or $r = 2$. Since f_k is a quarter period A_k is not half-canonical so $r = 1$.

If the g^1_4 had a fixed point it would have the fixed point counted twice since each A_k and B_k would contain the point, which we will call x. Thus the variable points in g^1_4 would be the hyperelliptic g^1_2 and the non-common points of A_k, B_k would be hyperelliptic Weierstrass points. Also the three distinct canonical divisors $A_k + B_k$ contain $2x$ and so $2x$ has index 2. Thus by the Riemann-Roch theorem x is also a hyperelliptic Weierstrass point and so each A_k and B_k is half-canonical, a contradiction. It follows that the g^1_4 must be without fixed points.

The g^1_4 defines W_3 as a four-sheeted cover of the Riemann sphere whose total ramification is twelve by the Riemann-Hurwitz formula. Each $2A_k$ and $2B_k$ contributes two or three to the total ramification depending on whether A_k (or B_k) is two distinct points or one point connected twice. But if the latter possibility ever occurred the total ramification would exceed twelve, so each A_k (and B_k) is two distinct points.

The covering given by the g^1_4 is thus ramified above six points on the Riemann sphere and above each such point there are two points of multiplicity two. But then the covering must be normal with the group of cover transformations being a four group since the monodromy group of the cover is the normal four-group in the symmetric group on four letters. There are two possibilities for $Z_2 \times Z_2$, namely, (3;0,1,2;0) or (3;1,1,1;0). But in the first case each fiber of the cover must be a canonical divisor, g^2_4, since it consists of the canonical g^1_2 doubled. Since the fiber defines a complete g^1_4 the second alternative must hold. q.e.d.

Combining Theorems 5 and 6 with the classical characterization of hyperellipticity for genus three gives the last two corollaries.

<u>Corollary 10</u>: Suppose a Riemann surface of genus three satisfies the hypotheses of Theorem 4, and suppose the theta function also vanishes to order two at a half-period. Then W_3 admits a (3;0,1,2;0).

<u>Corollary 11</u>: Suppose a Riemann surface of genus three satisfies the hypotheses of Theorem 5, and suppose the theta function also vanishes to order two at a half-period. Then W_3 admits a $Z_2 \times Z_2 \times Z_2$.

7. <u>Cyclic groups of order three for genus two</u>. We now prove the converse to Corollary 10, Part II.

<u>Theorem 7</u>: Let W_2 be a closed Riemann surface of genus two. Suppose there are two one-sixth periods f_1 and f_2 so that i) $\langle f_1\rangle \neq \langle f_2\rangle$; ii) $3f_1 \equiv 3f_2$; and iii) the theta function vanishes at f_1 and f_2. Then W_2 admits a cyclic group of order three.

<u>Proof</u>: By the Riemann vanishing theorem, let A_k and B_k be divisors of degree one solving the equations

$$u(A_k) + K \equiv f_k$$

$$u(B_k) + K \equiv -f_k$$

for $k = 1,2$. Then $3A_1 \equiv 3B_1 \equiv 3A_2 \equiv 3B_2$ by ii) and the four points are distinct by i). Thus $3A_1$ defines a complete g^1_3 on W_2 which represents W_2 as a three-sheeted cover of the Riemann sphere. The total ramification is eight by the Riemann-Hurwitz formula. Thus all the ramification occurs at the A's and B's and so each of the four branch points of the three-sheeted covering has multiplicity three. Consequently, the monodromy of the cover is a cyclic group of order three and so, therefore, is the group of cover transformations.

8. <u>Some local characterizations</u>. In all the characterizations discussed in this paper for genus five or three, the number of conditions imposed on the theta function turns out to equal the codimension in the appropriate Teichmueller space of the locus of surfaces admitting the particular automorphism groups. By viewing the theta functions evaluated at particular points of finite order as functions on Teichmueller space (or Torelli space or the Jacobian sublocus of the Siegel upper half space) these derived conditions give global defining equations for the subloci except that we must stay off the hyperelliptic locus for genus five. In this section we discuss cases where some derived vanishing properties give local rather than global characterizations; that is, we show that the loci of surfaces admitting particular automorphism groups are irreducible components of varieties defined by vanishing properties of the theta function. Two cases of genus five are considered. We will indicate the analysis for obtaining a local converse for Corollary 9, Part III and briefly discuss the case of surfaces admitting at least one fixed-point-free involution. The results of this section seem hardly definitive. One expects that the local equations are, in fact, global if we again avoid the hyperelliptic locus.

Let T be Teichmueller space for genus five. Let $(E - H)^k$, $k = 1,2,3,4$, stand for the loci in T of surfaces admitting k or more elliptic-hyperelliptic involutions. Then $(E - H)^1$ and the hyperelliptic locus of T are disjoint since a surface of genus five cannot be hyperelliptic and elliptic-hyperelliptic at the same time. Let (F) stand for the locus of surfaces admitting a $(5;1,3,3;1)$. By the Riemann-Hurwitz formula it is seen that the codimension of (F)

is eight in T and that of $(E - H)^3$ is nine.[4] Since every group of automorphisms generated by three elliptic-hyperelliptic involutions contains a (5;1,3,3;1) we see that $(E - H)^3 \subset (F)$. It is also known that each component of $(E - H)^3$ and (F) is a complex submanifold of T. Moreover, each component of (F) contains points of $(E - H)^3$.

For a particular half-integer theta-characteristic, consider $\theta[\eta](0;B)$ as a function on T. Thus for $W_0 \in (E - H)^3$ there are nine $[\eta_k]$ such that $\theta[\eta_k](0;B) = 0$, $k = 1,2, \ldots ,9$, are defining equations for $(E - H)^3$ at W_0. The $[\eta_k]$'s will be assumed to satisfy $\eta_1 + \eta_2 + \eta_7 + \eta_8 = 0$, $\eta_1 + \eta_3 + \eta_4 + \eta_9 = 0$ and $\eta_2 + \eta_5 + \eta_6 + \eta_9 = 0$, corresponding to the T-families for the three elliptic-hyperelliptic involutions of W_0. Thus $\theta[\eta_k](0;B) = 0$, $k = 1,2, \ldots ,8$ defines a pure four dimensional variety, V_4, at W_0. But a component $(F)_0$ of (F) passing thru W_0 lies in an irreducible component of V_4 at W_0 by Corollary 9.[5] Since $(F)_0$ is four-dimensional we see that this component of V_4 is $(F)_0$; that is, the eight conditions on the theta functions derived in Corollary 9 give local defining equations for $(F)_0$. Since any component of (F) contains points of $(E - H)^3$, the result is true for all of (F).

---

4) That this naive counting of dimensions is, in fact, correct is settled in Baily [5].

5) To prove this last statement precisely calls for much detailed analysis of the T-families associated with the three elliptic-hyperelliptic involutions which generate the elementary abelian group of order eight. We omit this analysis.

Let (FPF) denote the loci in T of surfaces admitting a fixed-point-free involution. Since any (5;1,1,3;0) contains a fixed-point-free involution, $(E - H)^2 \subset (FPF)$. Moreover, every component of (FPF) contains points of $(E - H)^2$. Let the defining equations at $W_0 \in (E - H)^2$ be given by $\theta[\eta_k](0;B) = 0$ $k = 0,1,2, \ldots ,6$, where $\eta_0 + \eta_1 + \eta_2 + \eta_3 = \eta_0 + \eta_4 + \eta_5 + \eta_6 = 0$. At W_0 let V_6 be the purely six-dimensional variety defined by $\theta[\eta_k](0;B) = 0$, $k = 1,2, \ldots ,6$. Since the component, $(FPF)_0$, of (FPF) passing through W_0 is six dimensional and lies in V_6, we see that $\theta[\eta_k](0;B) = 0$, $k = 1,2, \ldots ,6$ give locally defining equations for all of $(FPF)_0$, and so similar sets of six equations locally define all components of (FPF).[6)]

---

6) This last result was proved by D. Mumford (unpublished) by other methods.

9. <u>Closing remarks</u>. There are many obvious gaps in the preceding presentation. We mention several that seem of immediate importance. The first problem is to characterize elliptic-hyperelliptic surfaces of genus four. Any information on non-hyperelliptic automorphism groups for genus four would be a significant advance. Secondly, to finish the discussion of involutions on surfaces of genus five, the 2-hyperelliptic case remains. For all of the above cases, no reasonable conjecture exists. Thirdly, one would think that the strongly branched involutions $(p_1 > 4p_0 + 1)$ would admit a characterization for the cases not covered in Corollary 4. Finally, any characterization of groups of order three for $p \geq 3$ would seem to be a significant advance.

It should be remarked that the problem of characterizing hyperelliptic automorphism groups by properties of the theta function has been solved by classical methods quite different from those used in this paper.[7)] Moreover, Martens has shown how to distinguish hyperelliptic surfaces by vanishing properties of the theta function. Consequently, the problem of characterizing hyperelliptic automorphism groups by properties of the theta function is completely solved. A similar characterization of elliptic-hyperelliptic automorphism groups is an obvious next problem which seems to present interesting but not insurmountable difficulties.

Finally, we include a few remarks as to how the methods of this paper might yield information about non-abelian automorphism groups, at least for low genus. The surfaces of genus three admitting the

---

7) We know of no explicit reference; however, the methods occur in [25] and more recently in [17].

two largest automorphism groups we will denote W_{96} and W_{168} for the orders of the groups under consideration. Since neither of these surfaces is hyperelliptic, any four-group on them must be a (3;1,1,1;0). By consulting a table of the possible automorphism groups for a surface of genus three we see that the two automorphism groups are the only ones containing respectively seven and fourteen four-groups. By the characterization of Theorem 6 we see that the existence of seven or fourteen four-groups can be discovered immediately from the vanishing properties of the theta function.

A similar situation occurs when considering surfaces of genus three admitting two non-commuting involutions. For a non-hyperelliptic surface, these will generate a dihedral group of six or eight. Since the dihedral group of order eight contains two four-groups, which the dihedral group of order six does not, these two cases can be distinguished by vanishing properties of the theta function.

References

[1] Accola, R.D.M., Riemann surfaces with automorphism groups admitting partitions. Proceedings of the American Mathematical Society Vol. 21 (1969) pp. 477-482.

[2] Accola, R.D.M., Two theorems on Riemann surfaces with noncyclic automorphism groups. Proceedings of the American Mathematical Society Vol. 25 (1970) pp. 598-602.

[3] Accola, R.D.M., Strongly branched coverings of closed Riemann surfaces. Proceedings of the American Mathematical Society Vol. 26 (1970) pp. 315-322.

[4] Accola, R.D.M., Vanishing properties of theta functions for abelian covers of Riemann surfaces (unramified case). Advances in the theory of Riemann surfaces Princeton University Press (1971).

[5] Baily, W.L., Jr., On the automorphism group of a generic curve of genus > 2. Journal of Mathematics of Kyoto University Vol. 1 (1961/2) pp. 101-108, correction p. 325.

[6] Castelnuovo, G., Sur multipli du una serie lineare di gruppi di punti. Rendiconti del Circolo Matematico di Palermo Vol. VII (1893) pp. 89-110.

[7] Coolidge, J.L., A treatise on algebraic plane curves. Dover

[8] Edge, W.L., Three plane sextics and their automorphisms Canadian Journal of Mathematics Vol. XXI (1969) pp. 1263-77.

[9] Farkas, H.M., Automorphisms of compact Riemann surfaces and the vanishing of theta constants. Bulletin of the American Mathematical Society. Vol. 73 (1967) pp. 231-232.

[10] Farkas, H.M., "On the Schottky relation and its generalization to arbitrary genus," Annals of Mathematics, Vol. 92 (1970), pp. 57-86.

[11] Farkas, H.M., Period relations for hyperelliptic Riemann surfaces. Israel Journal of Mathematics Vol. 10 (1971) pp. 284-301.

[12] Farkas, H.M. and Rauch, H.E., "Two kinds of theta constants and period relations on a Riemann surface," Proceedings of the National Academy of Sciences, Vol. 62 (1969), pp. 679-686.

[13] Fay, J.D., Theta functions and Riemann surfaces. Lecture Notes in Mathematics 352. Springer Verlag, 1973.

[14] Humbert, G. Sur un complex remarquable de coniques et sur la surface du troisieme order. Journal Ecole Polytechnique Vol. 64 (1894) pp. 123-149.

[15] Krazer, A., Lehrbuch der Thetafunktionen. Chelsea

[16] Krazer, A. and Wirtinger, W., "Abelsche Funktionen und allgemeine Thetafunktionen," Encykl. Math. Wiss. II; B7, pp. 604-873.

[17] Lebowitz, A. Degeneration of Riemann surfaces. Thesis, Yeshiva University, 1965.

[18] Lewittes, J., "Riemann surfaces and the theta function," Acta Mathematica, Vol. 111 (1964), pp. 37-61.

[19] Maclachlan, C., Groups of automorphisms of compact Riemann surfaces. Thesis, University of Birmingham, England, 1966.

[20] Martens, H.H., Varieties of special divisors on a curve. II Journal für die reine und angewandt Mathematik. Vol. 233 (1968) pp. 89-100.

[21] Riemann, B., Gesammelte Mathematische Werke. Dover, 1953.

[22] Roth, P., Über elliptisch-hyperelliptische Funktionen. Monatshefte fur Mathematik und Physik. Vol. 23 (1912) pp. 106-160.

[23] Schottky, F., Über die Moduln der Thetafunktionen, Acta Mathematica, Vol. 27 (1903) pp. 235-288.

[24] Schottky, F. and Jung, H., Neue Satze über Symmetralfunktionen und die Abelsche Funktionen der Riemannsche Theorie. Sitzungsberichte Preuss. Akademie der Wissenschaften (1909) I pp. 282-297.

[25] Thomae, J., Beitrag zur Bestimmung von $\theta(0,0, \ldots ,0)$ durch die Klassenmoduln algebraischer Functionen. Journal für die reine und angewandt Mathematik. Vol. 107 (1870) pp. 201-222.

[26] Walker, R.J., Algebraic curves. Dover

[27] Wiman, A., Über die algebraischen Curven von den Geschlechtern p = 4, 5, and 6 welche eindeutige transformationen in sich besitzen. Svenska Vet. - Akad. Handlingar Bihang till Handlingar 21 (1895), afd 1, no. 3, 41 pp.

[28] Wirtinger, W., Untersuchungen über Thetafunktionen, B.G. Teubner, Leipzig, 1895.

Index

Abelian covers 12

admissible β 57

automorphism groups 2

- dimension of spaces admitting certain 63, 95
- elliptic-hyperelliptic 50, 52, 81, 86, 87, 88ff 95, 98
- genus 2 61-62, 94
- genus 3 53, 56, 59, 88ff, 98-99
- genus 5 53, 59, 79, 81, 95, 98
- hyperelliptic 31, 79, 98
- involutions (see p_0-hyperelliptic automorphism groups)
- p_0-hyperelliptic 50, 51, 52, 70, 74
- strongly branched 69
- unramified 17, 65

canonical divisor 7

Castelnuovo 66, 70

character 6

coverings

- abelian 12
- completely ramified abelian 40ff
- ramified 8
- strongly branched 69
- unramified abelian 65

half-canonical linear series 67

Humbert's surfaces 86

involutions (See automorphism groups, p_0-hyperelliptic)

Jacobian 1, 7

linear series 66

"p-2 conjecture" 67, 79

quotient genus 67

Riemann vanishing theorem 1, 7

strongly branched covering 69

T-family 83

theta function 2

transformation theory 36

Wirtinger 65

Notation

$\underline{b}$	analytic map	3, 4, 5
W/G	orbit space	3
A	abelian group	4
$J(W_1)$	Jacobian of W_1	4
u_0, u_1	maps of surfaces into Jacobians	4
M_0, M_1, M_A, M_{UA}	meromorphic function fields	4, 5
$\underline{a}$	homomorphism	5, 33, 34
$(\pi iE, B)$	period matrix	6
$\theta[\eta](u_j B)$	first order theta function with theta characteristic $[\eta]$	6
ν_j	multiplicity of branching	8
X_1, X_0	divisor of branch points and image under $\underline{b}$	8
K_0, K_1	vectors of Riemann constants	9, 10
c_j	$_j c_j$ 0 in $J(W_1)$	9
$n^{-1}u_1(\underline{a}Z_0)$		9
$_j^{-1}u_1(\underline{a}Z_0)$		9
e_0, e_1		10, 11
R	characters of the abelian group, G	12
χ_f, χ	characters in R	12
f_χ	meromorphic function associated with χ	12
V	a product of cyclic groups	14
$(\chi_1, \ldots, \chi_s)$	an element of V corresponding to χ	14

$(\beta_1, \beta_2, \ldots, \beta_s)$	an admissible element of V	15, 58
t_χ	an integer associated to χ	15, 16
ε_χ	1/n-period in $J(W_0)$	17
α	homomorphism of $H_1(W_0, Z)$	34, 35
Δ_0	a disk on W_0	41
Δ_1	$b^{-1}(\Delta_0) \subset W_0$	41
p_0-hyperelliptic		50
Z_2	group of order two	50, 51
[n]	generalized hyperelliptic theta characteristic	50
(a_j)	generalized hyperelliptic period characteristic	50, 51
$(p_1; p_2, p_3, p_4; p_0)$	symbol for a four-group on W_1	56
(σ)	period characteristic	68
$[\varepsilon]$	theta characteristic	68
g^r_n	linear series of dimension r and order n	70

Also see page 3.

Vol. 309: D. H. Sattinger, Topics in Stability and Bifurcation Theory. VI, 190 pages. 1973. DM 20,-

Vol. 310: B. Iversen, Generic Local Structure of the Morphisms in Commutative Algebra. IV, 108 pages. 1973. DM 18,-

Vol. 311: Conference on Commutative Algebra. Edited by J. W. Brewer and E. A. Rutter. VII, 251 pages. 1973. DM 24,-

Vol. 312: Symposium on Ordinary Differential Equations. Edited by W. A. Harris, Jr. and Y. Sibuya. VIII, 204 pages. 1973. DM 22,-

Vol. 313: K. Jörgens and J. Weidmann, Spectral Properties of Hamiltonian Operators. III, 140 pages. 1973. DM 18,-

Vol. 314: M. Deuring, Lectures on the Theory of Algebraic Functions of One Variable. VI, 151 pages. 1973. DM 18,-

Vol. 315: K. Bichteler, Integration Theory (with Special Attention to Vector Measures). VI, 357 pages. 1973. DM 29,-

Vol. 316: Symposium on Non-Well-Posed Problems and Logarithmic Convexity. Edited by R. J. Knops. V, 176 pages. 1973. DM 20,-

Vol. 317: Séminaire Bourbaki - vol. 1971/72. Exposés 400-417. IV, 361 pages. 1973. DM 29,-

Vol. 318: Recent Advances in Topological Dynamics. Edited by A. Beck. VIII, 285 pages. 1973. DM 27,-

Vol. 319: Conference on Group Theory. Edited by R. W. Gatterdam and K. W. Weston. V, 188 pages. 1973. DM 20,-

Vol. 320: Modular Functions of One Variable I. Edited by W. Kuyk. V, 195 pages. 1973. DM 20,-

Vol. 321: Séminaire de Probabilités VII. Edité par P. A. Meyer. VI, 322 pages. 1973. DM 29,-

Vol. 322: Nonlinear Problems in the Physical Sciences and Biology. Edited by I. Stakgold, D. D. Joseph and D. H. Sattinger. VIII, 357 pages. 1973. DM 29,-

Vol. 323: J. L. Lions, Perturbations Singulières dans les Problèmes aux Limites et en Contrôle Optimal. XII, 645 pages. 1973. DM 46,-

Vol. 324: K. Kreith, Oscillation Theory. VI, 109 pages. 1973. DM 18,-

Vol. 325: C.-C. Chou, La Transformation de Fourier Complexe et L'Equation de Convolution. IX, 137 pages. 1973. DM 18,-

Vol. 326: A. Robert, Elliptic Curves. VIII, 264 pages. 1973. DM 24,-

Vol. 327: E. Matlis, One-Dimensional Cohen-Macaulay Rings. XII, 157 pages. 1973. DM 20,-

Vol. 328: J. R. Büchi and D. Siefkes, The Monadic Second Order Theory of All Countable Ordinals. VI, 217 pages. 1973. DM 22,-

Vol. 329: W. Trebels, Multipliers for (C, α)-Bounded Fourier Expansions in Banach Spaces and Approximation Theory. VII, 103 pages. 1973. DM 18,-

Vol. 330: Proceedings of the Second Japan-USSR Symposium on Probability Theory. Edited by G. Maruyama and Yu. V. Prokhorov. VI, 550 pages. 1973. DM 40,-

Vol. 331: Summer School on Topological Vector Spaces. Edited by L. Waelbroeck. VI, 226 pages. 1973. DM 22,-

Vol. 332: Séminaire Pierre Lelong (Analyse) Année 1971-1972. V, 131 pages. 1973. DM 18,-

Vol. 333: Numerische, insbesondere approximationstheoretische Behandlung von Funktionalgleichungen. Herausgegeben von R. Ansorge und W. Törnig. VI, 296 Seiten. 1973. DM 27,-

Vol. 334: F. Schweiger, The Metrical Theory of Jacobi-Perron Algorithm. V, 111 pages. 1973. DM 18,-

Vol. 335: H. Huck, R. Roitzsch, U. Simon, W. Vortisch, R. Walden, B. Wegner und W. Wendland, Beweismethoden der Differentialgeometrie im Großen. IX, 159 Seiten. 1973. DM 20,-

Vol. 336: L'Analyse Harmonique dans le Domaine Complexe. Edité par E. J. Akutowicz. VIII, 169 pages. 1973. DM 20,-

Vol. 337: Cambridge Summer School in Mathematical Logic. Edited by A. R. D. Mathias and H. Rogers. IX, 660 pages. 1973. DM 46,-

Vol: 338: J. Lindenstrauss and L. Tzafriri, Classical Banach Spaces. IX, 243 pages. 1973. DM 24,-

Vol. 339: G. Kempf, F. Knudsen, D. Mumford and B. Saint-Donat, Toroidal Embeddings I. VIII, 209 pages. 1973. DM 22,-

Vol. 340: Groupes de Monodromie en Géométrie Algébrique. (SGA 7 II). Par P. Deligne et N. Katz. X, 438 pages. 1973. DM 44,-

Vol. 341: Algebraic K-Theory I, Higher K-Theories. Edited by H. Bass. XV, 335 pages. 1973. DM 29,-

Vol. 342: Algebraic K-Theory II, "Classical" Algebraic K-Theory, and Connections with Arithmetic. Edited by H. Bass. XV, 527 pages. 1973. DM 40,-

Vol. 343: Algebraic K-Theory III, Hermitian K-Theory and Geometric Applications. Edited by H. Bass. XV, 572 pages. 1973. DM 40,-

Vol. 344: A. S. Troelstra (Editor), Metamathematical Investigation of Intuitionistic Arithmetic and Analysis. XVII, 485 pages. 1973. DM 38,-

Vol. 345: Proceedings of a Conference on Operator Theory. Edited by P. A. Fillmore. VI, 228 pages. 1973. DM 22,-

Vol. 346: Fučík et al., Spectral Analysis of Nonlinear Operators. II, 287 pages. 1973. DM 26,-

Vol. 347: J. M. Boardman and R. M. Vogt, Homotopy Invariant Algebraic Structures on Topological Spaces. X, 257 pages. 1973. DM 24,-

Vol. 348: A. M. Mathai and R. K. Saxena, Generalized Hypergeometric Functions with Applications in Statistics and Physical Sciences. VII, 314 pages. 1973. DM 26,-

Vol. 349: Modular Functions of One Variable II. Edited by W. Kuyk and P. Deligne. V, 598 pages. 1973. DM 38,-

Vol. 350: Modular Functions of One Variable III. Edited by W. Kuyk and J.-P. Serre. V, 350 pages. 1973. DM 26,-

Vol. 351: H. Tachikawa, Quasi-Frobenius Rings and Generalizations. XI, 172 pages. 1973. DM 20,-

Vol. 352: J. D. Fay, Theta Functions on Riemann Surfaces. V, 137 pages. 1973. DM 18,-

Vol. 353: Proceedings of the Conference on Orders, Group Rings and Related Topics. Organized by J. S. Hsia, M. L. Madan and T. G. Ralley. X, 224 pages. 1973. DM 22,-

Vol. 354: K. J. Devlin, Aspects of Constructibility. XII, 240 pages. 1973. DM 24,-

Vol. 355: M. Sion, A Theory of Semigroup Valued Measures. V, 140 pages. 1973. DM 18,-

Vol. 356: W. L. J. van der Kallen, Infinitesimally Central-Extensions of Chevalley Groups. VII, 147 pages. 1973. DM 18,-

Vol. 357: W. Borho, P. Gabriel und R. Rentschler, Primideale in Einhüllenden auflösbarer Lie-Algebren. V, 182 Seiten. 1973. DM 20,-

Vol. 358: F. L. Williams, Tensor Products of Principal Series Representations. VI, 132 pages. 1973. DM 18,-

Vol. 359: U. Stammbach, Homology in Group Theory. VIII, 183 pages. 1973. DM 20,-

Vol. 360: W. J. Padgett and R. L. Taylor, Laws of Large Numbers for Normed Linear Spaces and Certain Fréchet Spaces. VI, 111 pages. 1973. DM 18,-

Vol. 361: J. W. Schutz, Foundations of Special Relativity: Kinematic Axioms for Minkowski Space Time. XX, 314 pages. 1973. DM 26,-

Vol. 362: Proceedings of the Conference on Numerical Solution of Ordinary Differential Equations. Edited by D. Bettis. VIII, 490 pages. 1974. DM 34,-

Vol. 363: Conference on the Numerical Solution of Differential Equations. Edited by G. A. Watson. IX, 221 pages. 1974. DM 20,-

Vol. 364: Proceedings on Infinite Dimensional Holomorphy. Edited by T. L. Hayden and T. J. Suffridge. VII, 212 pages. 1974. DM 20,-

Vol. 365: R. P. Gilbert, Constructive Methods for Elliptic Equations. VII, 397 pages. 1974. DM 26,-

Vol. 366: R. Steinberg, Conjugacy Classes in Algebraic Groups (Notes by V. V. Deodhar). VI, 159 pages. 1974. DM 18,-

Vol. 367: K. Langmann und W. Lütkebohmert, Cousinverteilungen und Fortsetzungssätze. VI, 151 Seiten. 1974. DM 16,-

Vol. 368: R. J. Milgram, Unstable Homotopy from the Stable Point of View. V, 109 pages. 1974. DM 16,-

Vol. 369: Victoria Symposium on Nonstandard Analysis. Edited by A. Hurd and P. Loeb. XVIII, 339 pages. 1974. DM 26,-

Vol. 370: B. Mazur and W. Messing, Universal Extensions and One Dimensional Crystalline Cohomology. VII, 134 pages. 1974. DM 16,-

Vol. 371: V. Poenaru, Analyse Différentielle. V, 228 pages. 1974. DM 20,-

Vol. 372: Proceedings of the Second International Conference on the Theory of Groups 1973. Edited by M. F. Newman. VII, 740 pages. 1974. DM 48,-

Vol. 373: A. E. R. Woodcock and T. Poston, A Geometrical Study of the Elementary Catastrophes. V, 257 pages. 1974. DM 22,-

Vol. 374: S. Yamamuro, Differential Calculus in Topological Linear Spaces. IV, 179 pages. 1974. DM 18,-

Vol. 375: Topology Conference 1973. Edited by R. F. Dickman Jr. and P. Fletcher. X, 283 pages. 1974. DM 24,-

Vol. 376: D. B. Osteyee and I. J. Good, Information, Weight of Evidence, the Singularity between Probability Measures and Signal Detection. XI, 156 pages. 1974. DM 16.-

Vol. 377: A. M. Fink, Almost Periodic Differential Equations. VIII, 336 pages. 1974. DM 26,-

Vol. 378: TOPO 72 - General Topology and its Applications. Proceedings 1972. Edited by R. Alò, R. W. Heath and J. Nagata. XIV, 651 pages. 1974. DM 50,-

Vol. 379: A. Badrikian et S. Chevet, Mesures Cylindriques, Espaces de Wiener et Fonctions Aléatoires Gaussiennes. X, 383 pages. 1974. DM 32,-

Vol. 380: M. Petrich, Rings and Semigroups. VIII, 182 pages. 1974. DM 18,-

Vol. 381: Séminaire de Probabilités VIII. Edité par P. A. Meyer. IX, 354 pages. 1974. DM 32,-

Vol. 382: J. H. van Lint, Combinatorial Theory Seminar Eindhoven University of Technology. VI, 131 pages. 1974. DM 18,-

Vol. 383: Séminaire Bourbaki - vol. 1972/73. Exposés 418-435 IV, 334 pages. 1974. DM 30,-

Vol. 384: Functional Analysis and Applications, Proceedings 1972. Edited by L. Nachbin. V, 270 pages. 1974. DM 22,-

Vol. 385: J. Douglas Jr. and T. Dupont, Collocation Methods for Parabolic Equations in a Single Space Variable (Based on C^1-Piecewise-Polynomial Spaces). V, 147 pages. 1974. DM 16,-

Vol. 386: J. Tits, Buildings of Spherical Type and Finite BN-Pairs. IX, 299 pages. 1974. DM 24,-

Vol. 387: C. P. Bruter, Eléments de la Théorie des Matroïdes. V, 138 pages. 1974. DM 18,-

Vol. 388: R. L. Lipsman, Group Representations. X, 166 pages. 1974. DM 20,-

Vol. 389: M.-A. Knus et M. Ojanguren, Théorie de la Descente et Algèbres d' Azumaya. IV, 163 pages. 1974. DM 20,-

Vol. 390: P. A. Meyer, P. Priouret et F. Spitzer, Ecole d'Eté de Probabilités de Saint-Flour III - 1973. Edité par A. Badrikian et P.-L. Hennequin. VIII, 189 pages. 1974. DM 20,-

Vol. 391: J. Gray, Formal Category Theory: Adjointness for 2-Categories. XII, 282 pages. 1974. DM 24,-

Vol. 392: Géométrie Différentielle, Colloque, Santiago de Compostela, Espagne 1972. Edité par E. Vidal. VI, 225 pages. 1974. DM 20,-

Vol. 393: G. Wassermann, Stability of Unfoldings. IX, 164 pages. 1974. DM 20,-

Vol. 394: W. M. Patterson 3rd, Iterative Methods for the Solution of a Linear Operator Equation in Hilbert Space - A Survey. III, 183 pages. 1974. DM 20,-

Vol. 395: Numerische Behandlung nichtlinearer Integrodifferential- und Differentialgleichungen. Tagung 1973. Herausgegeben von R. Ansorge und W. Törnig. VII, 313 Seiten. 1974. DM 28,-

Vol. 396: K. H. Hofmann, M. Mislove and A. Stralka, The Pontryagin Duality of Compact O-Dimensional Semilattices and its Applications. XVI, 122 pages. 1974. DM 18,-

Vol. 397: T. Yamada, The Schur Subgroup of the Brauer Group. V, 159 pages. 1974. DM 18,-

Vol. 398: Théories de l'Information, Actes des Rencontres de Marseille-Luminy, 1973. Edité par J. Kampé de Fériet et C. Picard. XII, 201 pages. 1974. DM 23,-

Vol. 399: Functional Analysis and its Applications, Proceedings 1973. Edited by H. G. Garnir, K. R. Unni and J. H. Williamson. XVII, 569 pages. 1974. DM 44,-

Vol. 400: A Crash Course on Kleinian Groups - San Francisco 1974. Edited by L. Bers and I. Kra. VII, 130 pages. 1974. DM 18,-

Vol. 401: F. Atiyah, Elliptic Operators and Compact Groups. V, 93 pages. 1974. DM 18,-

Vol. 402: M. Waldschmidt, Nombres Transcendants. VIII, 277 pages. 1974. DM 25,-

Vol. 403: Combinatorial Mathematics - Proceedings 1972. Edited by D. A. Holton. VIII, 148 pages. 1974. DM 18,-

Vol. 404: Théorie du Potentiel et Analyse Harmonique. Edité par J. Faraut. V, 245 pages. 1974. DM 25,-

Vol. 405: K. Devlin and H. Johnsbråten, The Souslin Problem. VIII, 132 pages. 1974. DM 18,-

Vol. 406: Graphs and Combinatorics - Proceedings 1973. Edited by R. A. Bari and F. Harary. VIII, 355 pages. 1974. DM 30,-

Vol. 407: P. Berthelot, Cohomologie Cristalline des Schémas de Caracteristique $p > o$. VIII, 598 pages. 1974. DM 44,-

Vol. 408: J. Wermer, Potential Theory. VIII, 146 pages. 1974. DM 18,-

Vol. 409: Fonctions de Plusieurs Variables Complexes, Séminaire François Norguet 1970-1973. XIII, 612 pages. 1974. DM 47,-

Vol. 410: Séminaire Pierre Lelong (Analyse) Année 1972-1973. VI, 181 pages. 1974. DM 18,-

Vol. 411: Hypergraph Seminar. Ohio State University, 1972. Edited by C. Berge and D. Ray-Chaudhuri. IX, 287 pages. 1974. DM 28,-

Vol. 412: Classification of Algebraic Varieties and Compact Complex Manifolds. Proceedings 1974. Edited by H. Popp. V, 333 pages. 1974. DM 30,-

Vol. 413: M. Bruneau, Variation Totale d'une Fonction. XIV, 332 pages. 1974. DM 30,-

Vol. 414: T. Kambayashi, M. Miyanishi and M. Takeuchi, Unipotent Algebraic Groups. VI, 165 pages. 1974. DM 20,-

Vol. 415: Ordinary and Partial Differential Equations, Proceedings of the Conference held at Dundee, 1974. XVII, 447 pages. 1974. DM 37,-

Vol. 416: M. E. Taylor, Pseudo Differential Operators. IV, 155 pages. 1974. DM 18,-

Vol. 417: H. H. Keller, Differential Calculus in Locally Convex Spaces. XVI, 131 pages. 1974. DM 18,-

Vol. 418: Localization in Group Theory and Homotopy Theory and Related Topics Battelle Seattle 1974 Seminar. Edited by P. J. Hilton. VI, 171 pages. 1974. DM 20,-

Vol. 419: Topics in Analysis - Proceedings 1970. Edited by O. E. Lehto, I. S. Louhivaara, and R. H. Nevanlinna. XIII, 391 pages. 1974. DM 35,-

Vol. 420: Category Seminar. Proceedings, Sydney Category Theory Seminar 1972/73. Edited by G. M. Kelly. VI, 375 pages. 1974. DM 32,-

Vol. 421: V. Poénaru, Groupes Discrets. VI, 216 pages. 1974. DM 23,-

Vol. 422: J.-M. Lemaire, Algèbres Connexes et Homologie des Espaces de Lacets. XIV, 133 pages. 1974. DM 23,-

Vol. 423: S. S. Abhyankar and A. M. Sathaye, Geometric Theory of Algebraic Space Curves. XIV, 302 pages. 1974. DM 28,-

Vol. 424: L. Weiss and J. Wolfowitz, Maximum Probability Estimators and Related Topics. V, 106 pages. 1974. DM 18,-

Vol. 425: P. R. Chernoff and J. E. Marsden, Properties of Infinite Dimensional Hamiltonian Systems. IV, 160 pages. 1974. DM 20,-

Vol. 426: M. L. Silverstein, Symmetric Markov Processes. IX, 287 pages. 1974. DM 28,-

Vol. 427: H. Omori, Infinite Dimensional Lie Transformation Groups. XII, 149 pages. 1974. DM 18,-

Vol. 428: Algebraic and Geometrical Methods in Topology, Proceedings 1973. Edited by L. F. McAuley. XI, 280 pages. 1974. DM 28,-